The Magic of Karma
Ancient Quantum Science?

And the *[bleeping]* Cause
of Religious & Political Cults??

By James T. Ordóñez

[The Angry Buddhist]

Inspired by
'*Biting an Iron Bull*' from Alan Watts[†1]
(†Highly recommended listen part 1 & 2 before reading)

The Magic of Karma
Ancient Quantum Science?

©2021 James T. Ordóñez
www.MAGICofKARMA.com
ISBN: 978-0-578-91006-2
First published in USA 2021
By Cosmic Wisdom Communications.

Cover Design by James T. Ordóñez
(Cover Background ©NASA / ESA/Hubble courtesy of NASA and STScI under contract NAS5-26555)
All illustrations, photos, and graphics, unless otherwise noted, are created by, and the property of the author, James Ordonez.
Production and typesetting by James T. Ordóñez
First Printed in USA

This book is designed to provide a comparison of political cults and religious cults with the alluded teachings of the Ancient Texts called the Puranas (Upanishads, Bhagavad Gita, etc.), for the sole purpose of exposing the inherent nature of avarice and racial undercurrent in all political and religious organizations and their leaders. It is sold under the express understanding that any decisions or actions you take as a result of reading this book must be based on your personal judgment and will be at your sole risk. The author will not be held responsible for the consequences of any actions and/or decisions taken as a result of any information given or recommendations made. If you got a problem with that, prove our point, please react with feedback —let's begin the *[bleeping]* conversation already.

Scan (or click) QR codes to visit website
(QR-Code images by QR Code Chimp)

Angry Buddhist Brand Logo design & production by the author, James Ordonez

DEDICATIONS

George Floyd (October 14, 1973 – May 25, 2020)
Heather Heyer (May 29, 1985 – August 12, 2017)

Greta Thunberg (Climate Change)
(Cowspiracy)
(The Green New Deal)

Abhai Charan De Bhaktivedanta S.
Alan Watts
&
All Our Teachers

[Got a problem with that?]

...because
one has to laugh at the outrage
to see...

Table of Contents

The Magic of Karma
Ancient Quantum Science?
And the [bleeping] Cause of Religious and Political Cults??

Please note:
The purpose of this book is by no means to denigrate but primarily to exalt and explain the hidden beauty of the Vedas, Puranas, and the ancient Hindu way of thought. Observations and personal experiences described by the author are an effort to correct what cults have twisted and left out overtime for their own financial and political agendas.
Got a problem with that too?
[Fuhgeddaboudit]

*No monks were harmed
with the writing of this work
...well, ...maybe a few inflated egos...
(ahankara)*

...

*A critical perspective
for students of Yoga
and Eastern Spirituality?*

Before Reading:

This entire Study is based explicitly on the sources;
therefore, pay special attention to all footnote references.

It would be best, but not essential, to have had a background
in Yoga philosophy or teachings from Ancient India.
This topic can get a little esoteric,
around the divine comedy, outrage,
and cult hypocrisies.

PLEASE DO NOT JOIN ANY CHURCH OR CULT.
That would defeat the purpose of these realizations!

If you're already in a cult, get the [bleep] out of there
as quickly as [bleeping] possible and don't [bleeping] look back.

Inspired by
'*Biting an Iron Bull*' from Alan Watts†²
†Highly recommended listen parts 1 & 2 before reading
http://bit.ly/Bite-This

ENJOY THIS RAW PURANIC PERSPECTIVE!

About the Author

[The Angry Buddhist]

James Ordóñez began studying Zen Buddhism in his early teenage years. He was a dysfunctional snot-nose thirteen-year-old and found a tattered book on a pigeon-pooped park bench—Alan Watts' "The Way of Zen."

In any case, there was something very wrong with Him.
There still is — Thank God!

Then, going into seventh grade, he was placed in an experimental class. They were lab rats, and 'Junior High School Quantum Theory' was the experiment. It was all downhill from there, into some rabbit hole that seemed to reflect his dark childhood—ironically. He began to understand why everything and everyone around him seemed selfish, almost cruel, and appeared to have little common sense and little, if no, empathy. He had to know what the *[bleep]* was going on in this lunatic fringe.

As for this book's subject matter, after studying this Ancient Science, he found the magic inspiring, yet dauntingly tragic, and *[bleeping]* comical (almost perverse)—in a divine sort of way that seems to explain modern religious and political cults, racism, and conspiracy theories! In the end, like it or not, all forms of discrimination, be they on skin color, birthplace, culture, accent, religion, belief, even political affiliations, are a form of racism driven by ill-conceived "ideas" from an inherent genetic and tribal nature, which are not going away—ever. The ˆanswer? Education on this Quantum-like science of Karma and Karma Yoga, and perhaps basic Quantum Science for all.

Without being outraged, there is no *[bleeping]* way to understand the outrage!

We hope you find this unique raw perspective both enlightening and enjoyable.

ॐ   ॐ   ॐ

Introduction

"If you're not outraged, you're not paying attention!" was Heather Heyer's last post before being run over and killed by hatred from a racial cult. Her sacrifice was one of the millions of observations of the previous several decades contributing to the need for this bird's eye view from Ancient Texts & Modern Science. The cause is more insidious than we can imagine. Without the outrage, there is no grasp of this topic—so we're breaking all the rules—yes—including grammar! Got a problem with that?

Guaranteed! A lot of folks are going to hate this—indeed, some cult members from both political and religious cults and just dull-headed racial conservatives living in a failed past. Ask us if we give a *[bleep]*. We don't!

We do apologize upfront for the level of detail; some may get a TMI headache. The whole picture cannot be stitched together without all the details and especially the *"outrage."* Without the outrage, there is no yin in the yang. The space inside a cup is the yin; the cup is the yang.

You've been warned. In fact, the welcomed negative and positive reactions may be our barometer that we may have hit a nerve. Throughout this study comparing the Ancient Texts with Modern Science, a poster child was needed to illustrate the similarities in behavior between political and religious cults, racism, conspiracy theories, and the inherent *(acquired)* nature of tribal, separatist outlooks, which foster and fester racism of all kinds.

The cult used willfully demonstrated, at least to us, their candidacy for *The Honor*—Poster Child. We do not apologize to any who disagree.

In the stewardship of progress, we hereby do our small part for the planet to help bring in a welcomed future of change and new tomorrows for personal self-care, introspection, and unique views for leaders and leadership—the combined perspectives of modern and ancient scientific thought, including DBT *(Dialectical Behavioral Therapy)*—better understood with this new lens.

Moreover, the theories proposed by Ancient Texts implore us to switch gears and to grapple with the herein discussed *dialectical* nature of existence, life, the cosmos, and why we are all so *[bleeped]* up—beyond repair without this raw Puranic *dialectical panorama*, which seems to echo modern science! This is both self-help as well as a 'save the *[bleeping]* planet perspective.

The only thing we guarantee is that upon finishing the book, a few essential perspectives may come back to haunt—good or bad.

– The Angry Buddhist

Glossary of Terms

*[bleep]***:** *The forbidden, taboo, cuss words! Insert expletive of your choice, i.e., F**k, S*ịt, Pị*s, etc. – Dialectical Therapeutic Reading allows the comedy of outrage as the lens of Emotional Intelligence Reasoning. i.e., "Because cult-ideologies are the [bleeping] Devil!"*

501(c)4: *The secret dialectical IRS political incentivized bribery for masked Corporate*
Oligarchic Totalitarian Control over "the 1%" weaponized by "Citizens United," upholding "Corporate Personhood—enslaving the world under minimum wage — AKA Runaway Capitalism."

Adhikar: *(Sanskrit) Coherence of spiritual desire depth & understanding.*

Ahankara: *(Sanskrit) "The temporary material ego." The first foreign substance to the soul.*

Akasha: *(Sanskrit) The deprecated "Lumineforous Aether vacuum of space. The second foreign substance to the soul.*

Batman: *The caped crusader—for [\$#ịts] & giggles.*

Buddha: *(Sanskrit) A human complete with Buddhi (knowledge).*

Bozo: *One devoid of Buddhi - a [bleep-head].*

Conservative: *One who conserves the failed past, another [bleep-head], afraid of progress.*

Dharma: *(Sanskrit) The inherent nature in each thing, creation, or development.*

Decoherence: *"Losely," the "Deeply Developing" argued, uncertainty Quantum Observation where waves of Consciousness collapse into matter—coherence.[3] This is where the Puranas allude to the entrance of life-force consciousness entering the material expanse, eliciting the cascade of Elemental Substances and plenums into measurable "Coherence."*

GOP: *The 'Good Old Boys' Grand Old Party Republican, "Conservative" "Bozo [bleep-heads]."*

Hermeneutics: *The study of idiotically accepting spiritual parables, lore, and legend as literal, which makes asses of all religions, destroying all that is Godly, perpetuating hatred and racism.*

Karma: *(Sanskrit) The web of action, propelled by Hegelian "thesis, antithesis, synthesis" as the unending web-chain of Cause and Effect.*

Maya: *(Sanskrit) "Measure" "Meter" "Matter" "Illussion"– The dialectical paradox of "Measuring" the infinite coheres the "24 Elemental Substances into the dialectical illusionary perception of matter through measuring-perception.*

Outrage Yoga: *The ability to see, through the lens of Outrage, Material Existence without being half-blinded and optimistic to the Samsara (Forest Fire) in a proven Dualistic, Yin Yang Universe(s).*

Puranas: *(Sanskrit) "The Ancient" [Texts], i.e. Upanishads, Vedas, Bhagavat Gita, Bhagavatam, Mahabharata, Ramayana, Sutras, etc., Controversially "All Ancient Hindu Texts."*

RatNut: *You'll have to figure this [bleeping] clown on your own when you read on and analogize him as the "Poster-Child" behind the comparisons with Political Cults.*

Sambandha: *(Sanskrit) Ancient Theory of Relativity.*

Shakti: *(Sanskrit) Energetic Consciousness fields, plenums, and waves, in Creation(s).*

Swami: *(Sanskrit) One who has "renounced" everything and should not have a single dime.*

Tao: *(Chinese) The unknowable source and principle guiding all reality. ("Tat" in Sanskrit, "That.")*

"The Cult:" *The poster-child religious, flat-earthers, cult example, led by "RatNut Swami."*

QAnon: *Hillbilly Rednecks believing in Jewish Space Lazers and amphibian pedophiliac cannibal extraterrestrials running the progressive political parties of the planet. (we wish this [$#¡†] were made up.)*

QOP: *The late GOP, may they "Rest In Peace." Next!!*

Quantum: *(Latin) "How much?" "Measure." The science of observing the infinitesimal to backward engineer coherence, in understanding "decoherence," where plenums come together and separate with the force of dialectically illusory Time -Space [measure]. So there!*

Vaishnava: *(Sanskrit) One devoted to the Vishnu form of God in the Hindu Puranas.*

Yoga: *(Sanskrit) "Yoke," to join, or connect.*

Zen: *The dialectical Behavioral Therapy for when we're taking religion too seriously literal, at face value, and all else is [bleeping] lost. Only go there if you've been hanging out with Bozos like RatNut Swami's Flat Earth Literalism or QAnon. Because then you need "major help."*

The truth is out there,
behind the laughter and the outrage,
and the clowns!

Note:

The Magic to this Madness
is in the authorized references
(Footnotes)

Scan the QR Code

(follow along)

Chapter 1

We write this to the sky above that it may rain below upon any who may thirst for something genuinely Cosmogonal.

Where to Begin

" *Since the elements (of creation) are derived from modes of 'sense perception,' they appear to be, in a sense, insubstantial and to be ultimately based on consciousness. Thus, from this Puranic perspective, the idea of decoupling an object from other matter and moving it through the ether[4] is much more plausible than it is from the standpoint of modern thinking."*

—*Quantum Physicist Richard L. Thompson, Ph.D.*

[Alrighty then!]

I n the end, the purpose of this book is to embrace the concept of *detachment* for holistic self-care. Ancient Hindu texts *(the Puranas)* explain this mystical science. These were written in ancient vernaculars and elaborate allusions, which, unfortunately, upstage some of the critical intended messages. The *Science of*

---

4   *The word 'ether' is used in this book to indicate the exhausted yet controversial scientific term for Luminiferous Aether (http://bit.ly/Luminiferous-Aether still under some controversy), the vacuum of space (or just "space"—in Sanskrit "akasha."*

Karma is one of those obscured. Volumes have been written on the subject. However, these are so enveloped in lore and legend *(for other lessons)* that the subjects of *karmic detachment* and *karmic entanglement* vanish into the proverbial "forest for the trees."

☞ *NOTE: 1.) We write this book in third person because the others involved are afraid to come forward, for fear of retaliation by the "Poster-Child" cult. 2.) The magic of this book is in the footnotes and authoritative science references linked therein to understand the entire compilation. We encourage the reader to assimilate all supporting 'footnotes' evidence carefully and continuously. 3.) And, if you haven't yet, the strongly recommended lecture by Alan Watts, Biting an Iron Bull (http://bit.ly/Bite-This), is a prerequisite to understanding the Quantum connections with Ancient Texts throughout all the references herein.*

On an individual level, understanding *karma* and its *karmic entanglements* become invaluable tools to navigate a healthy life and avoid toxic situations and people, even sometimes *detaching* from immediate family. Most cultures, however, place too much unconscious bias on the need to stay connected with unhealthy conditions, neglectful and abusive individuals, as in the subliminal "battered person syndrome."

On a global scale, communities, leaders, and countries would also benefit proportionally. Leaders might learn to lead by following, with accountability empathetically. Understanding *karma* identifies toxicity in their leadership and encourages *karmic detachment*, setting healthy boundaries for their communities and citizens' healthcare and welfare. Learning to detach and unravel, from negative *Karmic Entanglement*, is explained in these *Ancient Texts* as an essential healthcare practice for both material and spiritual life—*a Quantum era Education system may be in order.*

Admittedly, our initial outline for this book had sat around since 2013 before finding time from day jobs, or incentives, to get started truly. Having had negative experiences with some of the spiritual groups and cults, we associated with years back; we thought there were two separate books to write—one about the Ancient Science of Karma, the other an exposé on modern cults. Then came 2015, and a scary candidacy announced Donald J. Trump[†] for President of the United States. Something was terribly wrong—a gut feeling at first—and then we all began to listen. It became arguably apparent; a witless, uneducated, bigot, and con-man began receiving accolades and ratings to be The Leader of the Free World.

([†]As we found out, this theatrical clown is like a bad rash that won't go away!)

It was clearly a cult, a political cult! Those of us who have had experiences with cults were able to see. This time, however, it was not just a small offshoot of a religious *following* gone clandestine. An entire nation, the leading world economy for Freedom and Democracy, was held hostage by seventy million brainwashed cult followers seriously believing proven lie after lie, after proven lie, daily for six entire years. Many of us were praying that this Bozo and his deluded cult followers were just another temporary negative correction in the larger upward moving average of progress and progressive thought. But they just kept coming back, coddling and encouraging Nuclear Ready authoritarian dictators. *("Holy cow Batman!" we shuddered, "the crumbling world around us was about to be handed off to a babbling [bleeping] cult-leader!")*

Still, this *alerting* recent past did give us a much needed pause—a Karmic pause. As with typical religious cults, *cause and effect* from needed change had twisted and brainwashed half a nation for the worst. People join cults driven by dissatisfaction, seeking answers when change is much needed. Opportunistic cult leaders seeking power, money, and distinction take advantage, and a cult is born from folks seeking answers. This *action-and-reaction* moves a group, in this case, an entire nation, into potentially perilous circumstances. This political dip was clearly just another wake-up call. *"What next?"* we thought.

Still...
Was it ever so different?

As far as we can remember, we have had unending global wars that hover around the planet like plagues. Add to that recent Climate Change, global civil unrest, mass migrations, terrorist nations, genocides, overpopulation, mass homelessness, hundreds of thousands of essential species dying, polar ice-caps melting, and entire tropical forests burning. The whole ecosystem is a complete disaster, and life on the planet for the first time in Homo Sapiens history (*that's you*) appears to have no chance of survival. And corporations began assuming political power over 99% of the world's population! We were staring down the barrel of potential global domination by a U.S. President allegedly elected by foreign adversaries and Big Money and who very apparently didn't give a *[bleep]* about the global issues.

We were living through *history* in the making. During the COVID-19 Global Pandemic, worldwide civil unrest worsened; cults and conspiracy theories on the Internet were popping up everywhere, not only religious cults but political cults like QAnon. This one alleged that Earth is ruled "by a reptilian cabal of Satan-worshiping pedophiles." And it did not stop there. QAnon made its way into the United States GOP Republican Party with allegedly tens of millions of believers—recently dubbed the QOP. You can't make this *[bleep]* up if you tried.[5]

---

5   *New York Times – "QAnon," the new GOP – http://bit.ly/QAnon-the-New-GOP*
     NEW YORK TIMES – What Is QAnon, the Viral Pro-Trump Conspiracy Theory?"

("OMG," we thought,
"This is some bad [bleeping] GLOBAL karma!!")

This worsening world situation sparked a memory deep within. Karma is not limited to individuals but mostly collectively engulfs entire ecosystems, communities, cultures, entire countries, and yes, even a whole planet. We currently exist in a heightened state of urgency never before known to humankind. Ancient, age-old wisdom suggests a re-education for the fabric of human-ity's perception—an ancient *reprogramming* for survival. This book here proposes, after understanding these Ancient Texts, the reader may possibly agree that *collective karma* is driven by the *indiscriminate* herding, tribal, and survival instincts buried in our genetic memories. The Puranas allude to a catalyst of hope hovering in the ether around us—however, waiting to be let loose.

If only we can bring it out in the open.

Then, to top things off, the world began witnessing an explosion of conspiracy theories and cults beyond just religious and polit-ical.[6] Among these are "Chemtrails," "Black Helicopters," "the Bush Family, Bob Hope, and the Royal Family being extrater-restrials,"[7] and the most recent, a Republican Congresswoman

---

6   WikiPedia – http://bit.ly/ConspiracyTheories2020
     List of Conspiracy Theories 2020
7   TIME – The Reptilian Elite – http://bit.ly/The-Reptilian-Elite
     The Bush Family, Bob Hope, and the Royal Family are extraterrestrials

claiming that "Jewish extraterrestrials are shooting lasers from outer-space to set the forest-fires in California."[8] *(And we thought our premise for this book was far out there!)* And there are dozens more. Humankind is very susceptible to any nonsense that goes viral, especially these days on the Internet. Nevertheless, the indoctrinations into conspiracy theories are very much like that of cults and world religions. A series of *dialectical* ideas begin to be fed to a group. As the members of the group start to buy in, the momentum begins and credibility soars. The more the beliefs are accepted and repeated by others, the faster the indoctrination travels and spreads—like a virus. The Puranas hint at the logic behind that phenomenon as we have understood it from these Ancient Texts—maybe even with some *common sense.*

"Why all the [bleeping]?" you ask?
☞ *Let's face it; there are some emotional intelligence jargons that just cannot possibly be expressed by "normal" adjectives, politically correct colloquialisms, or any number of exclamation marks—in any language. Without speaking genuine gut-wrenching emotion, **without the outrage,** we will not be able to stitch all this together. Therefore, feel free to insert the expletive of your choice. You'll feel better and understand best if you do.*

We [bleeping] swear.

Some say these Ancient Puranic Texts began being recorded and passed on as far back as 25,000 years—vast through cultures and

----

8   *New York Magazine – Intelligencer – http://bit.ly/GOP-Jewish-Space-Lasers*
     GOP Congresswoman Blamed Wildfires on Secret Jewish Space Laser

time, extensive in content. So much so that one can propose almost any point of view, or skew, using the Vedas and Puranas. Those who have dabbled with *Vaishnavism* or *Buddhism* and studied these Texts (without a cult collective) will attest. No matter how outlandish an idea, with the expanse of *any* vast literature, one can convincingly propose a *position* on any topic. Similarly, the new QOP (the late GOP Political Party in the U.S.), or any conspiracy theory, can also argue any point no matter how outlandish with the available vast spectrum of rhetoric, news events, even scientific unrelated data. Like a game of telephone, *nuances spin themselves into a tangle.* They spin and twist ideas altogether dialectically to relate in the recipients' group mind—supported by other members of the group believing. And as with the Vedas, can argue any point. Such is the magic of *Maya's*[9] illusion that we see all-around in politics, cultures, conspiracy theories, and religions. It's the same old crap over and over, chewing the chewed.

    ☞ *Important, however, we have to warn you. Not everyone is going to get this quickly. We are all different and think very differently. For some, the notion of a "dialectical nature" will be common sense. Others will need to switch perspectives to a Bird's Eye View. Everyone will need to grasp the ancient teachings in some detail to stitch together this crucial ancient and modern Panorama. A new attitude is required— an Upside-Down, Inside-Out, and Backwards new perspective.*

----

9   *MAYA – The Shakti of Material Illusion – http://bit.ly/Basic-MAYA-Alan-Watts (the root word for matter, meter, and measure)*

The Puranas, and maybe some Quantum Science, suggest that like the physical world around us, conspiracy theories are merely dialectical projections, which become a reality to believers—somewhat unto a holographic universe projection principle.[10] Only, in this proposal from the Ancient Quantum-like Science of the Puranas, the perceptual projections are instead driven by thoughts and intentions (loosely instead of Quantum bits). This Dialectical nature in all things quickly makes staunch believers of willing, indoctrinated cult-members and conspiracy hate-groups in their group minds—no matter how outlandish or ridiculous an idea. Just look at all the insurmountable beliefs around the planet. They can't all be true.

...or, can they?!

The Puranic Karmic equation suggests that *thoughts* and *intentions* are on their own a *String* or *Particle* of some sort that together evolve into a synthesis—action or belief. Recent science around *Dialectical Behavioral Therapy*[11] loosely yet accurately encompasses this paradigm.

10   *VOX – Giant Hologram-Not-Far-Fetched* – http://bit.ly/VOX-Hologram-Not-Far-Fetched
     PBS - The Holographic Universe Explained – http://bit.ly/PBS-Holographic-Universe
     YouTube Hossenfelder Why Universe Hologram – http://bit.ly/Why-Universe-Hologram
     WIRED – 'Evidence' Hologram Universe – http://bit.ly/WIRED-Evidence-Universe-Hologram
     FUTURISM – A PHYSICS ILLUSION.– http://bit.ly/A-Physics-of-Illusion.
11   *Dialectical Behavioral Therapy* – http://bit.ly/Dialectical-Behavioral-Therapy
     DBT For Dummies In Fact, DBT For Dummies will help some wrap
     their brain around this illusive ancient Puranic Quantum Lens!

But we're getting way ahead of ourselves!!

With all this, the idea of two separate books became moot—one book on *Karma*, the other on the danger of cults. Combining the two made more sense, especially with the tragically comical similarities in behaviors between religious, *racial*, and political cults. A cult is a cult; and the U.S. QOP, QAnon Party *(the late Republican GOP)* of the United States had apparently become just that—a dark and insidious dangerous cult.[12] All political parties are *cultish*, but this one seemed to be waist-deep in an ugly past, racism, inequality, hatred, deceit, and supremacy, with constant spins on mistruths and manipulation of *"alternative facts."* They are blatantly and outwardly-open to living in proven failed past-decades and centuries, rather than embracing progressive thought, development, and new tomorrows.

"WTF," we thought, "these political Bozos are just like some of the Religious Cults we had encountered; irresponsible, dull-headed, and self-serving *only to their cause—at whatever cost to others."* Spreading their doctrine superseded humanities, civility, healthcare, ethics, even kindness, and empathy, just like religions. Just as suddenly, understanding basic *karma* for populations, governments, and the planet's survival became urgent and indis-

---

12   *The Republican GOP Death Cult – http://bit.ly/The-GOP-Death-Cult*
      Washington Post June 2020

pensable for general populations—a vital topic to ponder for future generations.

Interestingly, we rarely, if ever, hear the term *"racial cult."* There exists a kind of blind spot there—*denial.* Arguably, the word and effects of "racism" extend beyond race, color, nationality, and gender into all types of fearful unfair, discriminatory treatment of others due to mere differences in outward appearance, expression, or affiliation.

☞ *To be clear—for the remainder of this book—the word "Racism" (in modern-day synonymous with bigotry or prejudice) is not limited to race, nationality, or skin color, but as in real life, extends to all forms of separateness due to "supremacy" in either religion, belief, culture, economic class, gender, age, even cultural accents or mannerisms in communications.*

Like it or not, every form of discrimination or prejudice is a form of "supremacy." We need to be careful not to always equate this word *"supremacy"* with *"white."* The mere notion that one person has more rights than another is a form of supremacy—like it or not! No person ever truly believes that they are a racist, or supremacist, not even racist or supremacists' groups. There is always some justified defense. *"We are only protecting ourselves against the evils they bring to our lives and our families!"* saying in one manner or another.

This new perspective from the Puranas begs us to examine deeper. There dwells an inherent, inevitable, fearful racist, tribal default within all of us, without any exception—*observably*—like it or not! It's why we're here, in the material world to begin with *(according to the Puranas' logic)*! Whether the fear is of hue pigmentation in the skin, gender choice, geographic nationality, religious beliefs, dress, accents, or economic class, the cowardly fear of those who are different are all *"racism" and "supremacy"*—like it or not! When a parent insists that his child marries (or merely dates) ONLY within the same religion, nationality, or economic class, that parent is, to one degree or another, a "racist supremacist"—like it or not! The priest, monk, or swami, drawing boundaries between people based on arguable religious beliefs, is a *[bleeping]* "racist supremacist"—like it or not! To be clear—and we all may agree after a good grasp of this Puranic view—ALL cults, be they political or religious, are "racist" in nature and motive. All differentiation (*discrimination*), whether based on gender, skin color, age, beliefs, or class, like it or not, are synonymous with "racism." This dynamic nature of racism (*fear driven tribal separatism*) is not only inherited and built into our genetic memories from the values of past generations and past lives, but the Puranas give us another piece of the puzzle.

...

Moving on,

the ancient Texts (puranas) had given us clues

At first, we found it challenging to decide where to begin on this vital topic of *karma* and our primordial origins—as if all humankind were truly unaware of our very beginnings. According to the Ancient Texts, we may also all agree by the end of this Puranic report that *we are not entirely clueless of our origin and certainly not truly alone in the universe.*

The notion of having forgotten where we came from may merely be but a symptom of the very concept we now casually call *karma*. We had to ask, "where and how do we start this discussion?" We could have spearheaded this with philosophical arguments from various traditions in ethics. However, this little book's primary purpose is to keep it short, concise, and *light*. The very sublime point we stumble upon in these ancient teachings is precisely "just how absurdly simple" the entire mechanisms of life and creation are. Therefore, we decided to keep its *absurdity* and maintain the humor behind it all—*in the laughing Buddha style*—not taking it so seriously that it becomes impossible to achieve.

" *Bodhisattvas in Zen are often represented as bums ...the [Buddha] bum Hotei, who is always immensely fat. And he's saying, "Buddha*

*is dead ... I had a wonderful sleep and didn't even dream about
Confucius." And he's just stretching and yawning as he wakes up."*
Alan Watts 2.3.8 The World as Just So. †[13] *(See Note)*

Still, as you will see later on, this topic also becomes heady and
esoteric by default. It is a deep and obscure topic that we are not
used to pondering with our human conditioning.

Yet, it is tragically comical, and divine...

According to Ancient Texts, this *karmic* truth from the Puranas is
not only the simplest of explanations. It is also actually hidden
in plain sight—right under our very noses—literally. To what
universal end? We will explore this deep within some texts of the
Puranas and *Vedas* from ancient India.

☞ ***"Why was this book written in plural, third-person voice,"***
*you may ask? Over the years and decades, enough consensus
around this unique Puranic science perspective sprung up among
our peers and friends of the Yoga Community. At least two are
still tied to the "poster-child" cult in question and fear any
possible retribution to their livelihoods and families if exiled.*

---

13 † *HERE WE REMIND YOU: If you haven't yet, those wanting to best grasp the chapters
that follow, please consider listening to Alan Watts' "Biting an Iron Bull," our highly
recommended listen – part 1 & 2 before reading.
http://bit.ly/Bite-This*

Therefore, we agreed to represent the collective ideas as "The Angry Buddhist" brand and write the rest of this book in a collective plural voice. We also agreed that the inner collective voice we share, the companion in the heart of all living beings, had brought us together to explore these topics. The Angry Buddhist, therefore, remains a collective voice.

• • •

However, for me (*personally*), it began in early childhood. I grew up in a broken, abusive home, developing a bit of hypertension, awkwardly struggling to fit in, and pretty much alone, my family frequently moving. Then there was the racist world around us— never entirely fitting into a bigoted world; early school years were not fun. In early grammar school, I was almost albino white in a brown society. Then, at ten, America was our new home, I was now among white kids, but we did not speak the language, and were Latino. There was no lack of bullies in either scenario. On the other hand, one of my siblings was very brown and grew up compelled by cultural bias (even from immediate family) to be whiter somehow, *suffering* some need to be accepted as white. We all, therefore, turned out to be *[bleepholes]*, each in our own way—molded by societal pressures, from racism.

Still, I have to admit and feel fortunate, compared to others, I was just another peculiar child, alienated and removed, a little lost, always overcompensating, never entirely "fitting in." I still don't.

It seems many of us have less than perfect childhoods. Things like sports, competition, racism, war, and especially *folks eating animals* were the real lunatic issues for me growing up. I was a mere child. Yet, I saw the world and people, like I really did not belong here, as though I was born in this existence as part of some cruel joke. I suspect many folks have felt the same.

No worries, I will not go into self-serving nauseating details of my own idiosyncrasies. That, too, would defeat the purpose of this intentionally small read. I'll only bore you with a few brief pertinent instances to give a little perspective. I do have to declare to get it out of the way—as far as "sane" goes, I know that I am a tad loonier than most—actually a little crazy by *"norm"* standards. After all, I begin this book with quotes like "decoupling an object from other matter and moving it through the ether."

Sounds like the beginning of some science fiction novel? Yep! Crazier still may be that I believe this is both possible and potentially a common everyday occurrence all around us. We just cannot see it with our human faculties—suggest the *Ancient Puranas*.

I was just a little kid. The ancient Puranic Texts proposed, to me, a magical phenomenon that is simultaneously the very nature *(dharma)*, as well as a working function of the universe—karma *(mere "action")*. The mechanics explained there a simple primordial science, sustaining the balance of all things at all times.

"Ok, way out there," I thought, but I decided, "I would nibble." I was still in my early teens, and nothing else made sense anyhow, so I figured, *"why the [bleep] not!"*

• • •

One day, back at around five years old, I had planted a few white Lima-beans in the front garden of our house. In the days that followed, I woke up at sunrise and ran down to the front yard, excited to see the soil's magic of life appearing. One morning, each of the seven sprouts had a tiny white bean on top. It was amazing and magical! Every time I think of that morning, I remember the glee—my hair standing on end. Absolute unfor-gettable wonder about life came to me that morning with both a sense of relief and a feeling of actually belonging to something—of sorts!

As I held one of the newly sprouting white beans between my thumb and index finger, I noticed the lima-beans somehow resembled the end of my thumb in size, tone, and shape. With wonder and curiosity, I began staring closely at my finger, comparing the two. All in a sudden moment, I experienced my first childhood memory of an awakening epiphany. *We all have them; this was mine.*

Something both scary and dazzling happened. A daydream—I felt suddenly drawn in, and as if falling furiously, was pulled in abruptly deep into the tip of my thumb. It was dark and cavernous within, like the infinite night sky. I was not really frightened, rather in awe. All around, I could see tiny twinkling lights like stars revolving around other stars, and clusters of these everywhere, and clusters of clusters, in every direction, some clusters larger and closer, some distant and tiny. *(No, I was not tripping on LSD or Ayahuasca; I was five years old, and my parents were Conservatives!)*

"We are so small among the stars, so large against the sky." – Leonard Cohen
(©NASA / ESA/Hubble. Andromeda Image license courtesy of NASA and STScI under contract NAS5-26555.
On 'ePubs' & footnote link this image links to Hubble's top 100 celestial beauties[14])

That childhood gardening daydream seemed like some eternal distant memory eons ago. I actually felt I had been there before—a vague recollection of knowing what these clusters of light were—*magical worlds large and small intermingling and stretched across infinite microcosms and macrocosms*. Later I used to ask myself, "How the *[bleep]* did I even vaguely know that!"

---

14   Hubble Top 100 Images – http://bit.ly/Hubble-Top-100

Then came elementary school; I would see similar images revealing that my wacky fantasy had resembled solar systems or atoms, or both. Somehow in my dream, they were indeed identical—atoms and solar systems simultaneously existing as *one-and-the-same*— as I remembered the experience. *(Later in adulthood, this notion of "simultaneously one-and-the-same" snuck up on me again in the Puranas—strangely enough—as if kismet meant to be.)*

And, as swiftly as I was pulled inside this daydream, I was abruptly yanked out. As with a whoosh, I was back to squatting in awe on the lawn, comparing the white lima-bean sprout to my tiny thumb, in the small front garden of our home.

I ran inside to tell my family and house-keepers, but they all just stared at me blanked-eyed—and that was that. Future mentions elicited giggling and sarcasm, so I never brought it up again.

A few years later, in Fifth Grade at Catholic Church in New York City, Father Peter was teaching the daily 8:00 a.m. Catechism class to our group of altar boys. I remember three recurring concepts in his classes that would curl my insides even at that age.

"Animals have no soul" was the most disturbing by far because everyone around me was killing them and eating them as if there wasn't an abundance of other foods available and much less expensive. I would argue, "But they have families and feel pain!"

Coldly, he would reply in emotionless monotone, "Muscle reflex and instinct only."

Then, there was his constant reminder of the Ten Commandments that "you should not worship statues made of stone, or wood or marble," uttered in a church surrounded by marble statues of a Virgin Mary, St. Peter, and a host of angels. Folks would kneel with prayers and candles, especially at the bloody and ghastly violent wooden crucifixion. There were also twelve dioramas around the church depicting the "Twelve Stations of Christ," carrying his heavy cross and being whipped and speared on his way to his execution. *(That memory seems more like a LSD trip than my lima-bean hallucination)* Every time this commandment was discussed, I would raise my hand, pointing at the churchgoers offering candles and prayers to the various marble saints and God carved in wood, tortured and slain. *(I don't think the priest liked me very much. Thank God! [if you know what I mean])*

Father Peter and the other altar boys had learned to ignore me like I wasn't even there. I understood and couldn't blame them; I kind of felt sorry for them. The strangest teaching, however, which finally got me kicked out, was when Father Peter mentioned that "anyone who does not know about Christ and therefore does not accept Him as their Lord and Savior, will burn in Hell for all eternity." I raised my hand again, and begrudgingly was called upon, in monotone, "Yes, Jimmy?" "What about people that are born in China who never have a chance to hear about Christ? Do they burn in hell forever also?" "Yes, Jimmy!" Father Peter bluntly declared with certainty and commitment. I yelled out at the very top of my lungs, "Then God is very unfair!"

☞ *Like it or not, Father Peter and these Roman Catholic teachings were a specific form of racism (separateness), bigotry, and supremacy, being taught to small children not only the Altar Boys, but the entire school, and all Catholic schools, and other religious schools, all over the [bleeping] planet! Like it or not! And all religions do this! They teach the children the racism of separatism using "their" God—at a young age.*

Suffice to say, I was not allowed to come back to the altar boys, had to turn in my altar boy garments, and was politely asked not to come back to the Catholic Elementary School for the sixth grade. (*Hey, I could have ended up as another [bleeping] preyed upon statistic.*)

❝ *...the manifestation of racism as religious discrimination are often under-examined.[15]*❞
~Danielle Boaz – University of North Carolina

I was finished with wanting to be an altar boy, as was my grand-mother's wish that I become a priest in the Catholic Church. All

---

15   *The Color of Faith Discrimination – http://bit.ly/Religious-Racism
University of North Carolina*

that was gone, but not regrettably. What did not leave me with that expulsion was that I did not fit into this awkward world or with the people around me. Instead, what was reinforced was my need to understand why the *bloody hell* I was here, to begin with. "The world is mad," I thought, and I was only eleven years old. *(Talk about a [bleeped up] kid, hey?)* Much later, looking back, I realized the Catholic Church was just another cult, not just a religious cult, but a hybrid political, religious cult—"Religious Racism"[16]—discriminating even against other Christian cults.

Walking by a bench in Central Park later that Summer, someone had left a tattered book, *The Way Of Zen* by Alan Watts. *My personal journey began.* In the years that followed, although I couldn't really understand much at first, I became immersed in Alan Watts' book, and later *D.T.* Suzuki's Intro to Zen, and Roshi Philip Kapleau's The Three Pillars of Zen. I became a nerd and began meeting other nerd-teenagers. It just seemed like a way out.

A year later—as if by some preordained coincidence—I walked into the first day of my Seventh Grade Junior High School science class to hear the teacher declare, *"My name is Mr. Pepitone. You are in my class because you have been chosen for an experiment to learn a new science called Quantum Theory. This is a new*

---

16   *Religious Racism – http://bit.ly/Religious-Racism*
       University of North Carolina

science about the subatomic world, which explains how this blackboard eraser could actually be something as improbable as a cow." Someone yelled out from the very back of the room, "We want what you're smoking!" Laughing, he tossed the blackboard eraser at us. No, I did not catch it, I got chalked right on the head on day one, and it wasn't even me heckling.

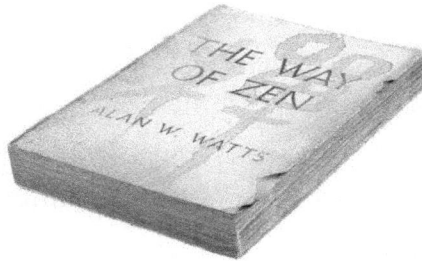

Because there are no coincidences!
(Photo, illustration & retouching by the author, James Ordonez using personal photos)

During these Middle School grades, I continued reading 'The Way Of Zen' and comparing with Mr. Pepitone's Quantum explanations of the subatomic universe—with a bunch of seventh graders. *(...because that wasn't already weird!)* We then began weaving in this new Quantum Science with Alan Watts' book. There were correlations and dialectical similarities for discussion with classmates and nerd friends.

I was still just a kid and had no true idea what was really going on, other than it all seemed to somehow tie in altogether—even going back to my five-year-old childhood lima-bean daydream hallucination. Junior High School became the beginning of a mentoring friendship with Mr. Pepitone's science class, Alan Watts' Way of Zen, and down the rabbit hole we went.

Later, at The High School of Art & Design, fellow students and even some faculty encouraged us to find spiritual teachers to bring it all together. Once again, I was kind of fortunate. This High School was one of the more progressive thinking, *avant-garde* schools in New York City. It was 1969, and both the student body and faculty were saturated in hippie culture and spirituality.

There were many Swamis and Gurus available, but one stood out, *Swami Bhaktivedanta*. That was when I also began reading *The Bhagavad Gita, The Upanishads, The Puranas, The Vedas, and The Ramayana.*

Compelled by dissatisfaction with the world and excited to understand, I dabbled with various traditions with teachings in Zen, Taoism, Tai-chi, Buddhism, and Vaishnavism. In the end, I became and now consider myself a combination of *Vaishnava Zen Quantum Buddhist*—I will explain later.

· · ·

☞ *Important Disclaimer: It was the late sixties and early seventies. The climate was perfect for finding fellow spiritual seekers and communities that thrive to this day. Sadly, however, most became corrupt organized religious cults, riddled with politics, criminal behavior, and abuse. Back then, these groups were blessings in the form of social vehicles for like-minded thinkers, spirituality, and truth seekers. However (and this is a big however), those groups and traditions are NOT what this book is about. Nor is this work about supporting or promoting any cults, political or spiritual. In fact, this book hopes to discourage the genetic tribal instinct to join cults because they always become the antithesis of free speech, individuality, and spirituality. The myth cults proselytize is that there is no possibility of "advancement" without "surrendering" one's Motivated-Reasoning[17] to their tribal leaders. Such tribal cults are the most dangerous spiritually and materially—impeding progress, ruining lives, and even families. As we unmistakenly found out with the Trump era, that is also very true of political cults. A cult is a cult, by any other name.*

So There!

As is typical with most organized religions, these modern cults eventually *fall* dark under the same spell they attribute to others. Many traditions consider non-followers as the *"unfortunate ones,"* who embody the very nemesis of the group's mission. Those to be *"saved"* (outsiders) are a devil of sorts as long as they do not comply and *"surrender."* Altogether, these groups and their members tend to go mad. To one degree or another, cult members

---

17   *Motivated Reasoning – http://bit.ly/Motivated-Reasoning*
      Psychology Today

live unhealthy lies and *distortions from dogma misconceptions*, totally skipping over their tradition's more profound messages.

Missing *the point*, they become raptured, mesmerized, and imprisoned by the *explicit* mundane face-value of the culture, language, and metaphors in what they read daily, rather than the *implicit* essence intended by their scriptures.

To "not follow and agree *wholeheartedly*" is taught an "ungodly and demonic" road to some hell or another—a kind of *the devil in others*. Our experience with these spiritual cults was as bad if not worse than our Catholic origins, only painted with big bright young smiles, flowers, flowery words, and folded hands. *(...and lots of "celibate" sex—more on that later)* Each tribe truly believes their racist sectarianism is "the *one and only* answer," and everyone else is evil or misguided. The mission is always to (*you guessed it*) exorcise the differences in others.

Aldous Huxley writes, *"The effects which follow too constant and intense a concentration upon evil are always disastrous. Those who crusade not for God in themselves, but against the devil in others, never succeed in making the world better, but leave it either as it was, or sometimes even perceptibly worse than it was before the crusade began. By thinking primarily of evil, we tend, however excellent our intentions, to create occasions for evil to manifest itself. No man can concentrate his attention upon evil, or even upon the idea of evil, and remain unaffected. To be more against the devil than for God is exceedingly dangerous. Every crusader is apt to go mad. He is haunted by the wickedness which he attributes to his enemies; it becomes in some sort a part of him."*

Have you ever been to a Multi-Faith gathering? It's like a convention of solipsists, each smirking behind poker faces, "*Mine is bigger than yours!*" They all get along on common grounds, but each of them knows at the core of their heart and soul that their version of the truth is far superior.

The hypocrisy of religions seems
to turn spirituality
into the proverbial racist Devil

Later, as an adult, I realized I was actually *"totally [bleeping] nuts"* only by the company I was keeping, i.e., cult members and fundamentalist extremists. There were all kinds—staunch followers, fringe communities, and mere sympathizers, all to varying degrees, some truly dangerous, others meek and kind. Most were indeed truly sincere and well-intentioned folks, but they sheepishly did whatever it took to fit the *cookie-cutter mold* provided by the cult. We did get to meet the Beatles' George Harrison in connection with the Hare Krishna group. Just a nice guy. He was, however, very straightforward with us, over the years, that he was not part of or supported *"the cult."* He would give some small donations as he did everywhere else. In the end, he was merely a *devotee of the Bhagavad Gita, the Vedas, and Puranas*—the ancient Hindu Scriptures—not *"the cult."* To this day, though, this group's knuckleheads still throw his name around as if he were a staunch follower—for their own proselytizing for donations—but that was never the case. *(They [bleeping] lie—period!)* The fact is, the Beatles subscribed to four controversial variations of Vedic Philosophy equally from four different gurus, without exclusive loyalty to one or another.[18]

George Harrison came to visit, albeit in complete secrecy. I received a call from Mukunda Das (*one of the good guys*), the head of Public Relations for our *"The Cult Branch"* poster-child. "We have a celebrity coming to visit, and I need someone *with hair* to guard him from the devotees." Many members felt that George Harrison should have surrendered everything to the cult as they had and given up all other religious associations in Hinduism. Around the same time, the famous *Swami Satchi-*

---

18   *The Beatles' Gurus – http://bit.ly/The-Beatles-Gurus*
      <u>*Vivekananda, Yogananda, Maharishi and Bhaktivedanta*</u>

dananda[19] (another renowned global guru, founder of Integral Yoga, and preceptor of Hindusim in the West) had visited. Upon arriving and leaving, he was verbally assaulted by many of the "Cult Branch" members, yelling and insulting him and calling him "impersonalist and Bogus," and other names. We watched in horror and embarrassment.

• • •

Cults and organized religions aside, we also realized from the Puranas the larger global insanity some of us see all around over our lifetimes, in politics, life, and spirituality. We, humankind, are, for the most part, lunatic narcissistic egomaniacs, tribal[20] to one degree or another. Some are drawn to extremes of anti-establishment mindsets, which often become tribal separatist groups, e.g., biker gangs, supremacists, spiritual cults, political cults, and even world religions. Each declares supremacy in one fashion or another.

Even crazier still is the inherent notion we all tend to believe from birth that we are indeed this material (flesh and bone machine) bodies we inhabit. That is the primary source of the world's lunacy.

19   Swami Satchidananda – https://bit.ly/SwamiSatchidananda
     Wikipedia
20   Tribalism Everywhere – http://bit.ly/Tribalism-Everywhere
     The Atlantic

That's like believing I am the car that I am driving! Even classical science gives us enough clues that our bodies, and those of other creatures, are machines inhabited by one form of consciousness or another. Yet, even mainstream religious and spiritual doctrines only emphasize ownership of the soul by the body rather than the other way around. The *Puranas* explain otherwise that consciousness *is* the *self*, or *soul*, the proprietor, which, in turn, inhabits a material vehicle that develops around the living entity based on *desire, intention*, and *thought* from *consciousness*.

But enough about our self-serving views on sanity, my petty childhood experiences, and my lunatic associations along the way. We simply felt the need to set the stage for how we arrived at writing this book. This work is about '*The Magic of Karma, the cult and racist phenomena,* and the similarities to Quantum Science. Let's now move on to the original teachings about *karma* from the ancient *Puranas and Vedas*.

...let's get this started!

...

In brief, our humble findings through a lifetime of seeking *meaning* have led us back full circle to the simplest of all fundamental truths from the *Puranas*. That is, and we summarize upfront, as follows:

All existence is an infinite ocean of dialectical consciousness in all directions, through all dimensions and spectra, as one single reality. In all life and all species, each living-being is but a mere drop of that cosmic ocean, both identical to the ocean, yet separate with free will—simultaneously—a droplet of consciousness in a vast and infinite ocean of consciousness, altogether dialectically creating all existences.

...but, that's not how it feels,
or where we appear to be!
Is it?!

☞ *...and you're probably wondering how any of this so far relates to Political cults. A cult is a cult, by any other name. Cults, be they political, religious, or the college fraternity, are all screwball unconscious herding instinct without empirical reasoning. They are born and sustained from the same thread, a dialectical "game of telephone"[21] inherent in humankind's tendency to distort and believe anything that is placed in front of us, as long as others agree (Confirmation Bias).[22]*

We only have to look and compare all the diverse, bizarre, and conflicting beliefs around the Globe to agree. They can't all be true. *Or can they?* The *Puranas* bring a new layer of explanation

---

21   *The Brain Distorts Accounts – http://bit.ly/Telephone-Games*
      <u>Northwestern University</u>
22   *Confirmation Bias – http://bit.ly/Confirmation-Bias*
      <u>Wikipedia</u>

to the nature of cults and conspiracy theories. The ancient Puranic science called *sambandha*[23] proposes that *everything*—without exception—comes from consciousness. Therefore, everything—without exception—is *dialectical* in foundation and behavior.

What the does that mean?? Read on. This theory becomes more clear in the chapters that follow as we dive into the mechanics of *sambandha*. We'll get back to political cults and wacko cults after studying the Puranic allusion of our *dialectical* existence from "*consciousness*" which causes all this mayhem—according to our takeaway from the Puranas.

☞ *REMINDER: If you haven't yet, those wanting to best grasp the chapters that follow, please consider listening to Alan Watts' "Biting an Iron Bull," – part 1 & 2 – our highly recommended listen (especially those not familiar with Eastern Philosophies) before reading. Behind the humor, these are esoteric metaphysics for which we will continue to remind you to "buckle up!" Go to: http://bit.ly/Bite-This*

----

23   *Sambandha – Relationship between all things – http://bit.ly/Sambandha-WikiPedia* *WikiPedia*

Migrations from the Ocean of Consciousness building the material universes
(Illustration collage by the author, James Ordonez)
(Background ©NASA / ESA/Hubble courtesy of NASA and STScI under contract NAS5-26555)

BUT FIRST!
...to grasp the connection
with cults, separatism, & racism
the 'Dialectical Foundation' needs to be
accepted!

Chapter 2

Who Am I and What on Earth Am I [bleeping] Doing Here?

...here Comes the Soul!

N ow repeat to yourself, *"I am simply just and only a single infinitesimal spark of consciousness, a droplet of mindfulness in an infinite ocean of consciousness and causality. As a drop in the ocean of consciousness, I represent the cause of all causes internally. Everything outside me, including this body I inhabit, is the effect of coming in contact with matter. While entangled with this matter, Cause-and-Effect is my relationship with both spiritual and material worlds. This Cosmic Relationship is my true yoga or union with everything, with all the universes, with God, and with all living beings."*

" Never was there a time when I did not exist, nor you, nor all these kings; nor in the future shall any of us cease to be."
~*Bhagavad Gita Purana 2:12*

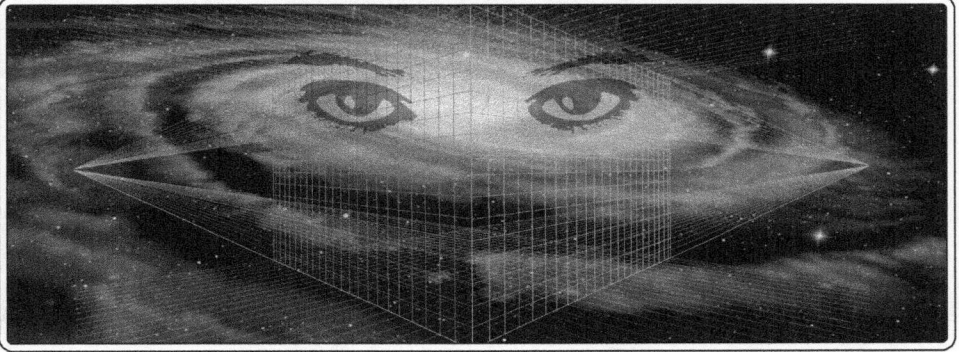

"The living spiritual being handles matter appropriately by his free-will and thus constructs his house. Similarly, matter is the ingredient only, but the spirit is the creator."
– *Easy Journey to Other Planets by Swami Bhaktivedanta*
(Illustration collage by the author, James Ordonez)
(background licensing iStock # 481229372.)

We begin the playful journey into consciousness

The Vedas and Puranas are the ancient Eastern Scriptures like the *Bhagavad Gita*, the *Mahabharata*, the *Ramayana*, and many *Upanishads*. Many say these Texts were written five to twenty thousand years ago. The format used is 'metered Sanskrit poetic verse' of mystical anecdotal narratives. Over time, these were often adorned or accompanied by colorful paintings and carvings. *These illustrative 'Ancient Texts' became influencers for many*

Eastern schools of thought and traditions, including Hinduism, Buddhism, Zen, and Taoism.

The Puranas attempt to illustrate that each energy or force in creation
is another wave of consciousness, a personage expanded from the ocean of consciousness.
Based on consciousness, each energy is personified (shaktis).[24]
(*Illustrations by the author, James Ordonez, from personal photos*)

Beyond the beauty of the illustrative anecdotes and art, we also find a splendorous hidden treasure—an ancient Quantum-*like* Science of Cause and Effect. *(Should we not give credit where credit is due?)* Dismally, these profound hidden messages are often left unseen to many students and even teachers—some unknowingly, others in strict conscious *censored* avoidance due to sectarianism, fears, dogmas, and even ungodly politics with "religious" leaders.

This ancient esoteric science is the *relationship* between all the Puranic sub-atomic forces and energies, also known as *shaktis*. The Merriam-Webster Dictionary describes *Shakti* as "*cosmic energy as conceived in Hindu thought.*"[25] Many sects in Hinduism

24   *Based on consciousness, each energy is personified – http://bit.ly/Hindu-Deities*
      WikiPedia – Hindu Deities
25   *Shakti – http://bit.ly/Meaning-of-Shakti – Literally, "Energy, ability, strength, effort, power, capability," the primordial cosmic energy and represents the dynamic forces that are*

also use the word *Shakti* to mean the Mother Goddess and female energy of the material world, also known as Mother *Durga*, or Mother *Kali*, or *Maya*. Other female personas are *Lakshmi*, *Parvati*, and *Sarasvati*, etc. All are manifest as "expansions" of the original female Goddess *Radha*, who without the male God *Krishna* cannot cope or even exist. (*Tragically comical is the fact that even with such a predominantly female cast among the Gods, misogyny somehow conveniently crept into many Hindu cults, no doubt by the mere misinterpretations we explore here.*)

" *Shakti: Energy, ability, strength, effort, power, capability; the primordial cosmic energy represents the dynamic forces that are thought to move through the entire universe"*
~WikiPedia

Much, much more profound, still, is the Puranic *Cosmic Relationship* between all these energies—*"sambandha."* In the *Vedas* and *Puranas,* these *relationships* are identified behind the anecdotal narratives.

Explicitly, at face value, these are all *deities* denominating various modes of worship. In many cases, these are religious factions claiming their deity is the first and foremost. Implicitly, however, it is explained in the Puranas that these cosmic energies are equal, various aspects, capacities, and constructs of the *Cosmic Whole.*

---

thought to move through the entire universe —WikiPedia
Merriam-Webster – Shakti

They assist in creating the material expanse. The different consciousness-based *shaktis* are represented by the *deities* for the narrative and vernacular used in human understanding and contemplation.

Even without the deities and paintings, these rhythmic, poetic Sanskrit verses from the Puranic Texts beautifully illustrate a deep Quantum-like-science of *karma, dharma*[26], and *yoga*—a paramount *Universal Relationship.* Everything encompassing this unique ancient thought is described there in exceptionally picturesque allusions. These are adorned in exaltation and portrayed in that universal vernacular of legend and lore so that we may understand, with our limited human experience, and to give them the reverence they deserve. *(Whoever said God does not have a sense of humor and imagination? How else would you try to get through to dull-headed primates taking themselves so [bleeping] seriously?)*

In essence, these Puranic verses are cumulative translations of a sort—*NOT to be taken literally in the material sense.* Sadly, most cults, religious fanatics, and fundamentalists insist that these legends are true and factual, *in our three-dimensional,* human-like corporeal manner. This is known as "Religious Liter-

---

26   *Dharma means the 'essential' inherent nature of any and all things – http://bit.ly/Meaning-of-Dharma*
      WikiPedia - Dharma

alism" in the studies of Hermeneutics. To that point, Dr. Denis Lamourex, Professor of Hermeneutics, from the University of Alberta comically sites as an example from Psalms 91:4, from the Old Testament of the Bible. *"He will cover you with his feathers, and under his wings, you will find refuge."* The professor then jokingly asks, "Is God [therefore] a cosmic Chicken?" It stands to reason that the biblical writer was being metaphorical in describing the act of taking God's shelter.

> ☞ *The common sense behind it all is simple and profound, but if you take it too seriously, you'll miss the point altogether. That's what the cults do—take themselves too seriously—and see what they end up achieving—racist segregation, abuse, pedophilia, crime, fighting, and even wars.*

Similarly, the illustrative anecdotal narratives throughout these Ancient Puranic Texts speak to us with metaphors. The Ultimate Truth is the essence of ubiquitous waves and atomic particles of consciousness—living beings called in Sanskrit *jiva* and/or *atman*. Everything is *personality* and *life-force*, pervading everywhere, encompassing everything down to the smallest particle. In other words, *Consciousness* is *Creationism*, in the proverbial nutshell.

The Puranas explain. *"The one" is also "the many."* In the infinite ocean of consciousness, a simultaneous symbiotic *relationship* is synonymously manifest between the sparks of consciousness and that infinite ocean itself. These phenomena occur concomitantly and simultaneously, playfully emerging in real-time, in the present eternal sacred moment. That cosmic *relationship* is called in Sanskrit *"sambandha"*—a word we should remember and will be touching on throughout this study. This is the Puranic version of a theory of relativity, which explains how things only have importance *"in relation to other things,"* which is the opposite of absolutism. Yes, this is similar to Einstein's Theory of Relativity, only playful from a dialectical perspective of ubiquitous youthful consciousness—*life-force*. This interactive *relationship (sambandha)* is precisely the main gist in these Ancient Puranic Texts.

Plenums of such consciousness are the substance(s) of all creation, as in the above-mentioned *shaktis*. The Puranas describe incorporeal personality(s) pervading everywhere, altogether creating the chaos and the order of everything. However, let's not be too quick to call them '*spirits,*' yet. Rather, think of them as forces, intentions, or impulses from *universal consciousness*.

Furthermore, these Vedic Puranas teach us that the secrets and truth about life, creation, and spirituality are merely insidiously concealed in the intricate web of material creation. They are simply hiding in plain sight, right under our very own sensory faculties—a comedic paradox—a veritable divine game of

'Hide-n-seek'—"*now you see it, now you don't.*" Yes, like a small child's game.

According to this ancient science, the Puranas' next conclusion is that the struggle between all things in this material expanse is about *'permanence'* versus the temporary material reality—*impermanence*. There exists no vernacular between this temporary material world and the permanent, eternal world-beyond. Of course not! There is no colloquial language, which could allow us to even fathom an *'other'* infinite reality. These eternal concepts can only be translated, or rather hinted at, with the only universal language that all humankind can understand—*lore and legend*.

However, to be clear, these Ancient Texts are not just some meaningless mythologies, as some atheists would think. Instead, these would be the closest possible translated illustrations (*allusions*), which many of us *inaccurately* label "baseless myth"—sadly! There exists no other colloquial language.

> ***And that is the paradoxical crux!***
> ***...from the Puranic point of view,***
> ***and among the verdicts of this study.***

Interestingly, T.S. Eliot (among others) played with this Puranic view in his widely regarded long poem, "The Waste Land"—a recommended read for poetry aficionados. The hidden nature of this cosmic message is described even there.

The Bhagavad Gita Purana reads, "This knowledge is the king of education, *the most secret of all secrets.*" When we look deep inside the messages within these Puranic illustrated anecdotes, we are indirectly instructed further. There is only one possible way to fathom, discuss, or understand *eternal concepts* with our human material *temporary* senses—what we may call meditative *allusions.* That meditative process is the reviving of our eternal *Relationship (Sambandha)* with the ocean of consciousness through the yoga of mindfulness and contemplative study.

Our limited scope and culturally conditioned faculties are incapable of seeing or understanding on their own. Spiritual *wherewithal (adhikar[27])* does not exist amidst material sense perception, spectra, of faculties to fathom anything non-material, or even the true meaning of infinity. We will discuss this *adhikar* (spiritual wherewithal) in-depth later, which is as essential as *sambandha* for this study.

This ancient science loosely encompasses *karma, dharma,* and *Yoga.* Since the only means for understanding the science are

---

27   *Adhikar – http://bit.ly/Adhikar-Qualification*
      Veda.WikiDot – Adhikar – In our context, a spiritual qualification and wherewithal that adorns and guides the individual drop of consciousness through the journey

these brilliantly illustrated translations, the same meditative mindfulness (*adhikar*) was needed to pass them down. The process was actually naturally lucid and organic. Collectively and over time, seers, sages, and monks received corroborating realizations from their meditations. They wrote them down in the best possible story-form for the consumption of the general populace.

The result was a funneling-down and *reductionism*[28] of magnificent messages. They describe unfathomable universal power, awe, and reverence, which become filtered down into acceptable human understanding—*color, form, personal human-like exchanges, and known symbols*. In the end, what is delivered to us is an illustration of unfathomable supernatural concepts—*translated* into poetic verse and lore.

A bit obsessed with the *Upanishads* from this Vedic literature, Carl Jung wrote, *"Mankind does not invent myths, it experiences them."* He adds, *"Indian philosophy is namely the interpretation given to the precise condition of the non-ego, which affects our personal psychology, however independent from us it remains. It sees the aim of human development as bringing about an approach to and connection between the specific nature*

---

28   *Reductionism*: – the practice of analyzing and describing a complex phenomenon in terms of phenomena that are held to represent a simpler or more fundamental level, especially when this is said to provide a sufficient explanation.

of the non-ego and the conscious ego. Tantra yoga then gives a representation of the condition and the developmental phases of this impersonality, as it itself in its own way produces the light of a higher supra-personal consciousness."— C.G. Jung[29] The Psychology of Kundalini Yoga. Another C.G. Jung recommended read is Man And His Symbols[30]. Jung's esoteric work sheds light allowing us to fathom mankind's dependency on symbology.

☞ *The secrets revealed behind the Vedic Puranas' illustrations are precisely this ancient science also described in Carl Jung's supra-personal consciousnesses (plural). They interact in playful dance and relationship, on a universal scale, with all other eternal ubiquitous consciousnesses—all together totaling the expanse of all existence.*

These Puranic Science are also some intense and esoteric topics that can only be grasped by looking deep inside the illustrated translations and then regularly meditating upon such depth within oneself—certainly not by the explicit face value of these mytho-logical illustrations alone. As do many religious cult members

---

29   *Carl Gustav Jung – http://bit.ly/Carl_Jung*
      Wikipedia – Carl Jung
30   *C.G. Jung – Man and His Symbols – http://bit.ly/Man-And-His-Symbols*
      Amazon.com – Man-His-Symbols-Carl-Jung

and teachings, those who remain adrift on the surface perpetuate their own illusion and grapple only with the mythology's mere *façade*.

Another Recommended Read & YouTube Channel
Reinventing the Sacred'

By Quantum Physicist Dr. Stuart Kaufman, Ph.D. [31]
(Photo & retouching by the author, James Ordonez from personal photos)

We realized while bringing all this lore together just how perfect and beautiful and harmonious the entire package of creation is. All this magical universe before us is godly, God-like, and entirely just and only God—altogether! Beyond the illustrated mythologies in the Puranas, a supreme consciousness lurks and rules an *order beyond the chaos*. From the spiritual perspective, Alan Watts says, *"Not a speck of dust in the universe is out of place."* Quantum Science seems to agree with a rather more complex approach.

---

31   *Reinventing the Sacred - by Stuart Kaufman - http://bit.ly/Reinventing-the-Sacred*
      Amazon – Reinventing-Sacred-Science-Reason-Religion
      YouTube – Stuart Kauffman Channel

> *...life exists at the edge of chaos ...It is almost spooky that such systems seem to coevolve ...The best exploration of an evolutionary space occurs at a kind of phase transition between order and disorder ... as if by an invisible hand, the system may tune itself to the poised edge of chaos..."*
> *~ Reinventing the Sacred*[32]
> *by Quantum Physicist Dr. Stuart Kaufman, Ph.D.*

Indeed, these Vedic Puranas' philosophy is the most colorful, beautiful, and personable we are yet to encounter among the world's many traditions in ethics and theosophy. Still, as mentioned, it is exceedingly important to note and mindfully remember that beneath the face value of the Vedic Puranas' stories lie the intended messages. Implicitly, they are arguably a quadrillion times deeper in essence and relevance than the face value attributed by religious literalism.

> *Myths are clues to our spiritual nature that help guide us to our sacred place within where we might unlock the creative power of our deeper unconscious self."*
> *~ Joseph Campbell*[33]

That is to say, do not get caught up in the stories and plots, thinking they are real in *this* realm of perception. They are not! And we will learn to argue that point to keep the balance! The plots and

---

32   *Stuart Kaufman, Ph.D. - Reinventing the Sacred – http://bit.ly/Reinventing-the-Sacred*
       Amazon.com – Reinventing-Sacred-Science
33   *Joseph Campbell — http://bit.ly/Joseph-Campbell*
       Wikipedia

characters in the stories are the best possible *representations* of magnificent consciousness-driven energies and forces behind the curtain of creation—*backstage if you will*. That is not to say there is no personality behind each of these Gods and Goddesses in the explicit representations. From our puny idea of *"person,"* human understanding doesn't even break the veil, especially against the infinite magnificence of the Personality in the *Summum Bonum*— the totality of everything—as alluded to by the *Puranas*.

And so, the *Puranas* continue to insist that these magnificent forces are indeed life-force, *supra-personal consciousnesses (plural)*, and worthy of awe and reverence. This also is part of the cosmic puzzle, and divine paradox, to carry and embrace in a mindfulness understanding.

We only speak here about the messages from these Vedic Puranas. However, it stands to reason that this mystic brain-teaser also applies to most other *so-called* mythologies and lore from other traditions. Sages and *thinkers* over the ages used allegories, semantics, and metaphors for delivering their *realizations*. How else could they address the general populace in all the world's scriptures to describe their visions and dreams of existences beyond this world? But what did the humans in charge do? They *[bleeped]* it up with Religious Literalism.

This confusion often causes intentional edits to the Texts for the aggrandizement or benefit of tribal cult leaders, religions, or even rulers—as in the case of King James[34]. We are reminded of the Tower of Babel from the Book of Genesis in the Bible's Old Testament. It was the time after the great flood. Logically, and according to some interpretations, the survivors agreed to build a high tower for shelter from another flood. One *thinker* had written:

> "Now the whole world had one language
> and a common speech..."
> *Genesis 11:1-2*

Shortly after, apparently, this was purported by someone to include that since the flood was sent from God, sheltering from God's natural disasters was "blasphemy." Therefore, apparently, a disclaimer was added.

> So The Lord said,
> let us confuse their language so they will not
> understand each other
> [and not be able to continue]."
> *Genesis 11:1-8*[35]

They made *Them* (God) out to be a *[bleeping]* Tyrant! Whatever the original intention, Biblical editors of the past turned the story

---

34   *King James version of the Bible – http://bit.ly/King-James-Bible*
      Wikipedia
35   *The Tower of Babel Text - http://bit.ly/Genesis-11_1-9*
      Genesis 11:1-9:

into a "punishment from God" for apparently humankind simply wanting to build a shelter from another flood and wanting to form community! You guessed it! So that the leaders could have the absolute power of the fear of God at their disposal to rule over others. Real nice guys!

Meanwhile, the likely original innocent anecdote behind the Tower of Babel was probably a simple message from thinkers and sages to not attempt to reach heaven, God, or spirituality through physical means.

Whether the Tower was actually finished is being debated by the Smithsonian with very thin allegations of proof.[36]   Still, it's no wonder why some religions and branches of many churches are still based predominantly on "the fear of God"—for political power.

Ancient or modern, humankind will not change its bigoted nature (*an "acquired dharma"*). Given the chance, "power corrupts and absolute power corrupts absolutely." Sadly, this trickles down *proportionately* all the way down and is not limited to kingdoms and governments, but through the fabric of enterprises, cultures, workplaces, organizations, and religions, even cliques and families. From the perspective of the Puranas, a *Modus Operandi*

---

36   *Smithsonian – Tower of Babel - http://bit.ly/Tower-of-Babel*
       Is the Tower of Babel real?

is built-in by default. A predatory nature is inherent in the cycle of birth and death. Not only the Puranas, but common sense tells us the survival instinct becomes a hunger of sorts prepared to gobble up everything in its path in a manic pandemonium stampede for survival—it's in the *genetic field* of the universe. From this phenomenon comes bigotry, racism, sectarianism, and boundaries of all sorts—to fear and identify differences in others that we will defend against and use for our own needs. It's a *[bleeping]* jungle out there, and religions and cults are never the exceptions, rather the justification. While to some, this may be common sense, others may need to be convinced.

> But once again,
> we're getting ahead of ourselves.

Coming back to the Puranas, it is exceedingly urgent to continually remember that, in essence, these Ancient Texts are *NOT baseless mythology.* They are the best possible translated illustrations *(allusions)* by seers, sages and thinkers, of quintessential and inexplicable realities beyond our grasp. As with Biblical anecdotes, we come to understand, from the Puranas, that this is true even in other traditions of theosophical philosophies where mythologies are the necessary vernacular.

> *All is a riddle and the key To the riddle is another riddle."*
>
> ~ *Ralph Waldo Emerson*

And so, our riddle continues even today. We use a *"poster-child"* example, of one such modern "Religious Literalism" cult. The classic problem in understanding *karma* lies with members of some Vedic cults and organized religious groups, ironically the very groups and Western *Gurus* who bring the *Puranas* to the Western World. This deficit is especially true of groups led by Western Swamis, like the ISKCON branch cult. We are all familiar with the saying, *"you can take the boy out of the mountain, but you can't take the mountain out of the boy."* These predominantly Western pseudo swami monks seem to have proven the above proverb to a "T," as you will see. Don't get us wrong, there are many good, well-intentioned folks in the group, and we introduce some of them later. But for the most part, many of their leaders who managed to climb up the corporate ladder ended up not being very nice folks at all. In fact, some are complete *[bleep-holes]*, lacking humanity, empathy, and civility, at any criminal cost to others, for the sake of *"the mission"* and the financial bottom-line—*their success.*

On the one hand, the cult members and leaders fall prey to the *same veil of illusion* described in the Puranas, which they sadly incessantly read daily without getting the point. On the other hand, due to nothing more than narcissistic human nature (and not getting the point), they become riddled with politics, money, and justified corrupt power. What trickles down to innocent cult members is a *cheating propensity*; likewise, individuals begin cheating, even in earning an "honest" living. *(We'll discuss that later as well.)*

> *If one associates with thieves, he will become a thief."*
>
> ~ *Bhaktivedanta Swami*

And, as if that wasn't enough, they all become experts at spinning and twisting the face value of the *Ancient Texts* and philosophies, at whatever cost to self-aggrandize their group above all other groups, in competition—like political cults. This often results in demonizing all other faiths, groups, or non-believers with fear and disdain—even very identically thinking groups. A form of racism ensues, "Those who are not like us are lesser, less fortunate, not in the graces of our God, can never possibly go to heaven, a devil of sorts." But, it's all for competition and money. This sectarian (political) competitiveness is the travesty of organized religions and cults. The Puranas' intended messages, *such as this science of Karma*, become entirely covered over and unavailable.

The result is *material-intention* with flowery words, smiles, and folded hands. Behind the *façade* lurks skewed vision and materialistic consciousness in just about every word, agenda, or deed. Such behavior is often criminal and abusive in order to defend its mission. A verse from the Bhagavad Gita (9:30) is commonly used to justify misdeeds and even crimes by the cult leaders. *"Even if one commits the most abominable actions, if he is engaged in devotional service, he is to be considered saintly because he is properly situated."* According to the cult interpretations of this verse, if someone is in good standing with the cult, they cannot be held responsible for unethical or criminal behavior. Hence, their teachings become tainted with malignant justification. We include a chapter of true stories that will curdle your blood and recommend reading with caution—not for the squeamish.

With these smudges blemishing the group's blistering consciousness, the majority are unable to see beyond the Purana's picturesque illustrations and are blinded to the secrets behind them—the proverbial forest for the trees. The followers and leaders can only see the face value of the mythological vernacular and become mere fundamentalist extremists pitying, shunning, or even fearing non-followers. This is where organized religion becomes the proverbial Devil.

This misled group, *our poster-child*, the above mentioned Hare Krishna "Cult branch," currently led primarily by *"His Holiness"*[37]

---

37   *The reader will need to assess this "Holiness" title on their own after assembling the entire picture herein.*

RatNut Swami† and his oligarchy of predominantly misogynistic male leaders, have managed to conclude from reading the Puranas at explicit face value that the Earth is flat.[38] *(You read correct, "Flat-Earthers!!")* We can't make this stuff up; please Google it. *(†This cult leader too, like Trump, also is a bad rash that won't go away—as you will see!)*

☞ *Please note that this ISKCON cult branch and "Mr. RatNut Swami" DO NOT represent other Hare Krishna Vaishnava groups. To avoid confusion and focus on the true intention of this book, we will henceforth refer to the branch as "The Cult" branch, and this fraud clown as "Mr. Swami."*

Together, these two will be our "Poster Child" example for this study on the nature of all types of cults in general: "Mr. Swami" and "The Cult" Branch.

They are convinced, on such religious literalism, that the Earth is a physical Frisbee-shaped disc. Somehow members have accepted that the Sun rotates around the Earth and that the Moon is further away than the Sun—physically! The group has reportedly spent over one hundred million dollars of donation money collected to

---

38   This links to the "Way Back Machine Web Archive" and may take half a minute to load. www.krishna.org/the-flat-earth/
...and they are all fighting about it all over the web– http://bit.ly/ISKCON-Flat-Earth-Controversy
Modern day ISKCON cult fighting over Flat Earth

build a gargantuan West Bengal Temple to the Physical Flat Earth Doctrine. *(Half of those donations were reportedly embezzled, with violence, and riots among the group members, and allegedly the burning of local homes, by local mafia, for not selling their land to build the planetarium. Our sources for these allegations, although hearsay, are reliable enough that we would welcome an investigation.)*

The Flat Earth?

There are many things we do know as facts from *Srimad-Bhagavatam*. As far as the earth *Srimad-Bhagavatam* does not describe it as a ball spinning in space.

But we can say for sure on the strength of the authority of *Srimad-Bhagavatam* the earth is not a globe spinning in space...

So there is no ball earth floating in space in the *Bhagavatam*.

The cult has applied "Religious Literalism" to anecdotal explanations of plenums and forces mentioned in the Puranas (shaktis). They conclude the Earth is a flat Frisbee-like disc, with the Sun physically rotating around the Earth closer in physical distance than the Moon.

(Illustration by the author, James Ordonez – Earth map adapted from free Pixabay image on https://www.pexels.com/photo/black-textile-41949/)

When one reads the Puranas *correctly,* the embedded messages speak loudly that these references to Earth, Sun, and Moon are distinctive "plenums of consciousness" that are *karmically* intertwined on an "ethereal" *(sub-atomic)* level alone—not physicality. This is a perfect example of the before mentioned "hermeneutical conjecture" (Religious Literalism) as with Psalms 91.4. This clarity is astounding for one who reads the Puranas correctly. What is even more astonishing, even alarming, is how an entire group can extrapolate and believe the *Sun orbiting a flat Earth, the same sun being closer to Earth than the moon*, at explicit face value.

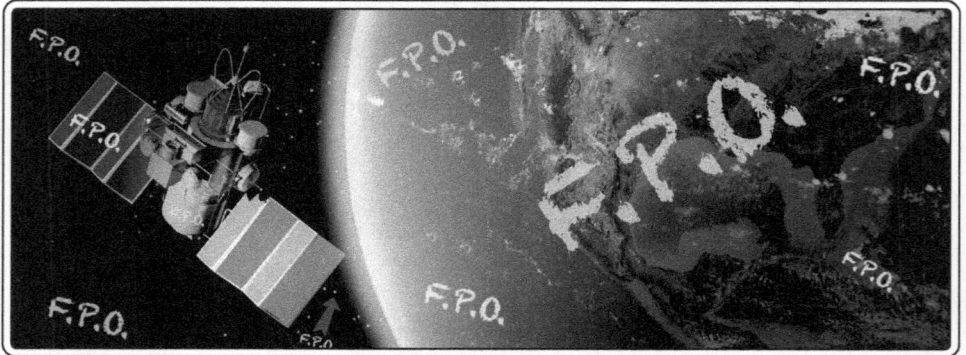

MEANWHILE: "Flat-Earthers" argue Earth space-photos an elaborate conspiracy by NASA, Russia, and China—altogether—in elaborate, costly image creation campaign to hide the flat earth reality from the public. There is even squabbling going on about this among some of "The Cult" factions:[39]
F.P.O. = "For Position Only," in Graphic Design creation lingo.
(Illustration collage by the author, James Ordonez – backgrounds licensing iStock # 611215490)

How then can they possibly believe this flat-earth doctrine?

---

39   NOTE: This Way Back Machine Web Archive may take a minute to load,
      Go smoke a joint.
      www.krishna.org/the-flat-earth/

For the same exact reason, some extremist fundamentalist cults believe that seventy-two virgins are awaiting them on the other side if they kill nonbelievers.

Rather than seeing the ancient science revealed behind the *representations*, the cult members become fundamentalist extremists. They revel in believing that all others not accepting cent-percent the face value of their *interpretations* are the very antithesis of goodness and godliness. Those nemeses are the "unfortunate" and "fallen," to be "saved." They are considered sort of a contaminated enemy to be wary of—if not converted. This separatism is clearly as evident in all types of cults—political, religious, or wacko spin-offs like QAnon. Leaders of "The Cult" in question, like other cults, likewise, behave less than righteous—at whosoever cost for the sake of the mission. It's their nature.

☞ *This separatist posture gives us a stark revelation with a simple **back-door-secret** to truly stay detached and clearheaded. That is, be sure to completely and without question avoid the association of ALL members from these organized groups. Cult knuckleheads remain riddled with politics, judgment, money, and deception. Such association is contagious **and as addictive** as gambling or drugs.*

According to these Puranic teachings, everything in the universe(s) is *based* on consciousness. We become supra-impressionable to the state of consciousnesses and misconceptions of those around us, like osmosis of sort. The secret here is to avoid these groups and cults like the proverbial plague.

Bhaktivedanta Swami, often used to emphatically preach, *"If you associate with thieves, you will become a thief."* After his passing in 1977, the politics of power and money turned the majority of their leaders into thieves. Of course, crimes and even abuses are justified using "the Mission" as the excuse. Instead of leaving altogether, the majority of the well-intentioned yet sheepish followers go along, even begrudgingly. The herding instinct takes over, and we tend to become captive in groups and cults. An unconscious bias runs through the group consciousness, dictating that there is nowhere else to go—*codependency*. The victims of this group's psychological captivity become addicted and find it difficult, even impossible, to leave, like the battered spouse syndrome.

I, too, personally, got sucked in and quickly became disillusioned and left on three different occasions when associating with *"The Cult"* branch of the Vaishnava groups. The first time was curiosity during my college days. That interest only lasted a few years. I left in light of corroborating reports of the cults' criminal behavior, including murder, child sexual abuse, and allegations of bodies buried in West Virginia hills—where *"Mr. Swami"* was second in-charge. We will be sharing only a few of the hundreds of stories later on. Nevertheless, cult members and even visitors are lectured from the Bhagavad Gita's verse, as mentioned earlier (9:30). *"Even if one commits the most abominable actions, if he is engaged in devotional service, he is to be considered saintly because he is properly situated."* This use of scripture to justify crime and abuse is not only enabling behavior but complicit.

Hastily, we ran the other way and pursued career and college. Much later, we attempted to bring our professional services to help their elderly with care, hospice, retirement, and social security. Still, we were shunned by "the leader(s)" and prominent members who became *defiant and outright hostile* for attempting philanthropy at three of their different locations. More on that at the end of the book.

Our advice upfront? If you have already joined a cult, please *leave*!! Run as *[bleeping]* fast as you can and take with you only what sound wisdom you may have found there—your heart will tell you. Study *with this new critical perspective in mind*. Practice on your own, and don't look back. Those who never have joined a cult may read and immerse themselves in as many "Texts" as they can digest. Look *within* daily while reading and meditating, and at all times, for inner guidance. This is where the supra-personal consciousness truly exists for each of us—not in the association of cults, their twisted lies, and less than honorable behaviors.

Enough about these types of religious cults, *for now*. The juicy details that matter to those wanting to join or leave these cults will be described later—when a better understanding of the real *Puranic Science* is explained. Let's talk about *karma, plenums of consciousness*, and the Quantum-Like Science embedded in the Vedic Puranas.

We promised back in chapter one to revisit the Puranic explanation of modern cults and Conspiracy Theories as we move along. The ancient Puranic science called *sambandha* proposes that everything—without exception—comes from *consciousness*. Therefore, everything—without exception—is *dialectical in foundation and behavior*. We asked, "What the [*bleep*] does that mean??"

This means "That *everything-everything* is a tangle of thoughts, ideas, and desires from an infinite number of *lost* living beings all together dictating the entire manifest world around us." (*Because... that's not a scary Sci-Fi horror "space-flick!*)

☞ *Ok! You ask, "What are we smoking??"*

In other words, yes, from one perspective, we are all *[bleeping]* crazy in a lunatic fringe of creation, like some *Sci-Fi* insane penal colony in outer space. In a nutshell, this Puranic explanation of the *dialectical consciousness foundation* is why human beings are so impressionable to believe and follow any crazy theory stitched together with random ideas, faulty interpretations, or just lies and deceit. Thoughts, ideas, and untruths are what drives us all while in this material existence. Selective Perception[40] is often the case, from Confirmation Bias[41] or motivated reasoning, or both, creating Cognitive-Dissonance-Behavior.[42] We are all, in fact, lost, period! And, we tend to cling on to anything that has a following—political cults, religious cults, conspiracy theories, or social media *[Bull-$#¡t]* if it has enough *"likes."* We're not the brightest bulbs in the universe, as we like to think.

But all is not lost! *Silver linings do exist in the plot.* It's still essential to understand *"why"* and how to detach from the proverbial asylum around us. A good grasp of this Quantum-like ancient Dialectical Science, new quantum science, along with knowledge of *Dialectical Behavioral Detachment,* will harness

40   *Selective Perception – http://bit.ly/Selective-Perception*
      Wikipedia
41   *Confirmation Bias – http://bit.ly/Confirmation-Bias*
      Wikipedia
42   *Cognitive-Dissonance-Behavior – http://bit.ly/Cognitive-Dissonance-Behavior*
      Simply Psychology

enough wherewithal (*adhikar*) to navigate this cult-driven world. A new Education System may be in order—one that supersedes Classical Physics and sectarian religious schooling with modern and ancient Quantum Thought. The latter, of course, religious sectarianism in schools, a tough nut to crack.

Read on. This becomes more transparent in the chapters that follow, as we unravel into the mechanics of *Puranic sambandha* and understanding the Puranic allusion of our *dialectical existence*, the illusion from "consciousness," which causes all this mayhem—*us, you-and-I*.

Just Remember!
Don't join any cults,
political or religious,
It's NOT the way home

Shiva Nataraja the Destroyer - Engaged in the Dance of Time [43]
(Illustration composite by the author, James Ordonez, from personal photos)

---

43   *Shiva Nataraja – http://bit.ly/Shiva-Nataraja*
      Ancient.com – Nataraja – The Dance of Time.

Chapter 3

Oh, Really? So, Where is Home?

...and how the [bleep] *did I get here?*

> *Jung believed that myths and dreams were expressions of the collective unconscious, in that they express core ideas that are part of the human species as a whole. In other words, myths express wisdom that has been encoded in all humans, perhaps by means of evolution or through some spiritual process."*
>
> ~*WikiVersity Jungian Psychology*[44]

As we have discussed from the Puranas, the instruments for these cosmic and divine revelations are but visceral *translations* from another dimension, way beyond the web of life and creation. Seers and introspects across human history have translated what they unraveled in their meditations into one single vernacular. That is the set of Humankind's symbols that can be

---

44   *Classical Mythology/Jungian Psychology – http://bit.ly/Jungian-Psychology* *WikiVersity*

discerned and read inherently by human experience and point of reference. This is the agency we like to call mythology.

The Puranas illustrate. Hiding in plain sight, this simple Truth of *ubiquitous consciousness* is the most insidious, ever-escaping, subtlest of all illusory magic for the living being. This magical plenum is also, in essence, the very foundation of all life, of all living beings, and all material existence. Therefore, the Puranas' conclusions are that the fabric of reality is in actuality *universal consciousnesses (plural).*

These same 'Texts' confirm that this knowledge is so simple that it is, therefore, *the* most challenging thing to see, or ponder, or comprehend with our human limited senses and faculties. This truth is simply hidden in plain sight. As mentioned before, the Bhagavad Gita reads, *"This knowledge is the King of Education, the most Secret of all Secrets."* – (BG 9:1) [45]

...and this is among the major conclusions in the Puranas!

It all sounds crazy at first, doesn't it? It actually is! ...even from the point of view of the Puranas! These ancient 'Texts' explain the

45   Recommended reading Free Online – http://bit.ly/The-Bhagavad-Gita-As-It-Is
      1972 Edition of 'The Bhagavad Gita As It Is'. Keeping in mind the
      recommendations from this work, The Magic Of Karma An Ancient Quantum
      Science. In other words, DO NOT join the [bleeping] cult in question!

absurdity of it all and how we are all trapped in a temporary yet potentially unending vast web of *cause and effect—imprisoned in a matter of speaking.* This is precisely what our book is all about. We are looking to explain the Puranic illusion and the spell behind the magic of all material creation, as taught by these ancient writings. The Sanskrit word for this spiritual madness is *Maya*— the root of the English words, 'measure' and 'matter' and 'meter' and 'mother'—yet originally meaning literally *"measurable, illusion, that which is not eternal [which can be measured]."*

Since there is no easy place to begin, let's just start with *you and I, and us, and now*—the present moment.

For this meditation and journey, we *(you and I)* therefore need to forget everything we know, everything we have learned, everything we have experienced, and our view of the world. We must now remind ourselves daily and moment by moment that 'this is a brand new lens.' Everything that has been taught to us over the years from our childhood, by our parents, friends, and society, must be put aside wholly to fully grasp this new—yet ancient— *optic.* We need to learn to see the world from a uniquely new angle of vision. Actually, we like to describe this unique perspective as *backwards, upside-down, and inside-out.*

• • •

Are you ready? Buckle up!

The first step is to accept or imagine that we are not our bodies, but that our bodies are simply vehicles, like an automobile.

"I" ride within!

...

This ancient *Quantum-Like Science* alludes that our bodies are a complex labyrinth, computer-like network of *dialectical, consciousness-based intentions and reactions*, preprogrammed by thoughts, desires, genetic memories, and past activity—*karma*—*and held together by 24 elemental substances—plenums.*

We *(you and I)* need to literally clear our 'cache' (yes, like a computer)—all of it—at least just long enough to have a discussion, read this book and bring this new perspective into our daily meditation. Ideally, we aim to share this *new perspective* and *"lens"* as your very own subjective meditation, your optional outlook on life, and your fresh way of thinking—your very own *mindfulness* to sculpt around your life experience.

Let's begin with this simple truth that most of us may agree upon at this point. "We are not these bodies that we inhabit, *we are actually consciousness*." The Sanskrit words for these individual drops of consciousness are *jiva* (the living spark) and *atma* (the soul)—together *jiva-atman*.

Most of us are intuitive enough, aware, or at least well-read sufficient to fathom or even agree on this point. Some cannot unravel the idea, and that's okay; the Puranas explain why. The ability to see, or not, is that *spiritual wherewithal* we mentioned before that in Sanskrit is called *adhikar*.

The lack of this *adhikar* quality is not necessarily a flaw or fault. There is no 'better or worse' or scale to compare. Instead, *adhikar* is developed by the desire and intention to awaken one's true spiritual constitutional position and true self beyond these material bodies. To not do so is merely a matter of choice and free will. We will discuss this in-depth later.

...a drop in an ocean, a wisp of consciousness

The spark of life, the *jiva-atman,* belonging to the infinite ocean of consciousness, shares the same inherent nature of also being infinite and without end. The Puranas explain that the energy that composes the *jiva-atman* transmigrates from body to body in different life forms across and throughout the various universes.

The Bhagavad Gita reminds us in the second chapter, *"Never was there a time when I did not exist, nor you, nor all these kings; nor in the future shall any of us cease to be."* – Bhagavad Gita Purana 2:12

The illuminations that are translated in the Puranas and Vedas speak of our journey through multiple universes and life forms. You and I, all of us—creatures large and small—are traveling through this vast sojourn exploring free will. We are mere sparks of consciousness each on our own journey of curiosity—an expansive excursion and wandering across time and space from one universe to another seeking some answer.

"This knowledge is the most secret of all secrets", only hiding in plain sight.

One of our spiritual teachers, *Swami Bhaktivedanta,* used to say, "This knowledge is simple for the simple and difficult for the difficult." In other words, there is a reasonable probability that this new *lens* will only work for those open enough and humble enough—therefore capable enough—to put aside preconceptions. We go about this journey *backwards, upside-down and inside-out,* down into the proverbial rabbit hole.

This journey begins not just before matter, but before thought...

We start our sojourn long before time, before impermanence and decay, before ether and sound, distance and *measurement (Maya)*, positive and negative, before independence, before *I-Me-Mine*—before creation itself.

The ancient Puranas tell us that before the material creation and all material manifestations, the only existence and only reality pervading all infinity is a pure unadulterated infinite vibration of 'loving' creative consciousnesses *(plural)*.

This vibration is described there as a serene yet dynamic universal self-awareness—*"personality"*—in playful indulgence with *them-self*, the all-pervading "cause of all causes," the ultimate "complete whole." This all-pervading cognizance is composed of an infinite number of sparks of pure unadulterated consciousnesses *(plural)* in all directions.

Just as an ocean is ultimately composed of water droplets, these infinitesimal living sparks of consciousness altogether form the ocean of consciousness. Their only end-game is to sustain harmony and maintain cohesion for the *Complete Whole* of all realities—*that same infinite ocean of consciousness.*

☞ *You're probably asking, "How is all this the simplest explanation? And what does this have to do with the nature of cults and racism?" As esoteric as this may sound, we promise this does come back around full circle to a straightforward and transparent understanding—once we can see the holistic bird's eye view stitched altogether with dharma, sambandha, karma, and all the puranic shaktis.*

Patience, grasshopper!

To our human ears, this *all-pervasive vibration* would appear to be a vast, eerie silence—an unending abyss of darkness to our eyes and an infinite, lonely emptiness to the rest of our physical faculties. It would be as if nothing were going on backstage behind our 5% perceivable creation.[46]

However, within the material universes, this plenum of eternal vibration can be compared to the modern subatomic fields,[47] which permeate both Luminiferous Ether[48] in the vacuum of space and the energetic composition of all matter. The Puranas describe this all-pervading vibrating field of consciousness as having an infinite multifarious variety of unlimited possibilities in infinite diversity.

These ancient texts describe an ever-expanding consciousness and

---

46   The "5% perceivable Universe" – http://bit.ly/Dark-Energy-Matter
      Wikipedia – 95% of the universe appears to be invisible
47   The Higgs Field – http://bit.ly/Higgs-Boson-Field-Mechanism
      The Higgs Boson Field Mechanism
48   One and the same?
      Controversy of Aether vs. Quantum fields – http://bit.ly/Aether-vs-Quantum-controversy
      Physics Exchange

intelligence pervading all directions beyond time and space in an ongoing playful vacillating union and separation—oscillating, reverberating and resonating forever, an everlasting dance.

Rasa Lila – The Celestial Dance – © Brooklyn Museum – WikiMedia Commons
(Image license courtesy of WikiMedia Commons from The Brooklyn Museum.
"The Brooklyn Museum can find no known copyrights" at the printing of this book. See Footnote) [49]

The Puranas *describe* such playfulness and dance include innumerable forms and *personas* manifested from consciousness—the foundation of all thought. At the center of the revelry whorls a *"Divine Couple"* (the cause of all causes) leading an eternal dance with their consorts. All these forms delight in unending revelry, interactive exchanges, and personal relations. These ancient texts implore us to *take a quantum leap of faith* through our new lens

- - - - - - - - - - - - - - - -

49   *Rasa Lila - The Celestial Dance – https://bit.ly/Rasa-Lila*
      Brooklyn Museum—WikiMedia

and try to fathom this unique and separate reality—not a *heaven*, but a looking glass.

We continue our meditation; take a deep breath, open your mind even wider, and now open your heart.

What follows is not understood
with the intellect alone

While inconceivable, we must *try* to imagine a place of eternal, multidimensional, diverse, pure intention, and playful, *"loving"* relations and festivity among an infinite number of limitless beings—the true meaning of infinity.

Generally, we think of infinity in terms of space and time. However, the Puranas teach us that actual infinity exists beyond infinite space and infinite time, *in endless dimensions of consciousness*—a paradox to ponder upon, like a Zen Buddhist Koan[50] meditation.

...and down the rabbit hole we go!

The ancient texts explain that this realm manifests from *and within* a vast, ever-expanding consciousness, harmoniously diverse *energetic intelligence* driven by *reciprocal interaction*—a cosmic coherence. Imagine the constitution of this realm being

50   Koan Meditation – http://bit.ly/Koan-Meditation
         Wikipedia

composed of pure goodness, balance, and rhythm within every intention, every purpose, and every deed.

Our human limited understanding and experience could only describe this serene expanse as some heaven of sorts. Imagine an unadulterated, ever-expanding, energetic force of pure good and loving interactions—a primordial substance of infinite symphonic cooperation. While some may wish to call this Heaven, others may even also call this concept God.

However, the ancient texts of the Puranas are, in fact, still only describing the expanse of an *"effulgence,"* an infinite field of radiation where the confluence of two streams converge and struggle—*independent free-will* versus *dependent full surrender.'* This endless radiation expanse is called in Sanskrit the *Brahma-jyoti*, the effulgence of *Brahmán* (God).

For this work about the Magic of Karma, we will not go into the topic of surrender to God, or Heaven as a destination. That begins to fall under the category of theism and religiosity, not what this Karma reflection is about. We will dive into that topic in future editions of this work as they develop. Karma is only inherent with those *jiva-atmans* who choose the path of independent *free-will* by coming into the Material Expanse. Here we begin that Puranic exploration of our decision and entrance into this world.

...now take a really deep breath!

Our journey began within the energetic field or effulgence mentioned above. As per these Ancient Texts, the struggle that initially manifests for the living being, the *jiva-atman*, starts there. This living entity, a spark of life, is faced with the most critical decision of its entire infinite existence—*individuality or collective existence*—a paradox. That *decision maker's* harmonious nature and consciousness are already equally both individuality yet collective intelligence, simultaneously. *How does one decide? Indeed!*

The Puranas teach that the *jiva-atman* individual spark of life—a wisp of consciousness—with the ocean of consciousness are simultaneously the same and different, one and yet separate. This phenomenon is called, in Sanskrit, *achintya bheda abheda.*[51] This paradoxical Relationship—being qualitatively the same yet completely different—is the magic behind the curtain. Therefore, this paradox is also essential in understanding the universal rule of the *relationship* between everything, which we described before as *Sambandha*. Like the drops of water in an ocean sharing oceanic water quality, the *jiva-atman* shares all the qualities of the entire ocean of consciousness.

----

51   *Simultaneously One and Different – http://bit.ly/Same-and-Different*
        <u>*Wikipedia*</u>

Beyond this decision to explore *free-will* and individuality lies full immersion, *Vaikuntha* (the Sanskrit term literally meaning 'that realm free of exhaustion and free of *karma*')—heaven. As mentioned, this topic also begins a dive into the bottomless abyss of arguments in theism. Our theme on *karma* is not to be confused with theosophy or religion.

...

The *Relationship* to remember always (the main principle in *sambandha*) is that every atomic particle, speck of dust, or spark of life in existence is "*one yet separate*" and "*same yet different*" from the totality of this ocean of consciousness. The paradox continues.

The Puranas explain, they *always travel together*. The creation and the creator, the source, and the effect, the *energy*, and the *energetic*, are always inseparable. This companion, the *energetic*, the *source* and *creator*, is called in Sanskrit the *Param-Atman* or Super Soul—much like the Holy Ghost in the Catholic tradition of the Trinity. The Super-Soul, however, is described as "*Simultaneously One and Different*" with the embodied *jiva-atman* living entity—in all qualities.

Your fork in the road...

Imagine yourself as a living photon particle of light in flight, within and as part of an all-pervading expanding radiance, like a thrusting solar wind traveling between and outside infinite sphere-like multiple universes. Try to picture it and ponder—this alone is a meditation.

It is a timeless journey of blissful tranquility in an undeterred, undefined symmetry of belonging and affinity. Then, from within an endowment of *free-will*, along comes a craving curiosity, an irresistible urge to choose, like a tantalizing breeze out of nowhere — the first-ever independent meandering in your experience. Like a fork in the road, a sobering yet perplexing option appears. "This way or that?" "To continue united or disperse?" "To give or to receive?" "To accept or to explore?" "To the realm of *Vaikuntha* or to the unknown?"

The difficult choice is to either fully immerse and belong to the primordial center and *cause of all causes* in continued reciprocity with infinite possibilities of harmonious *Relationship (sambandha)*, ...*or* to indulge in an alternate subjective individuality alone through independent exploration—an unknown exciting adventure.

An *innocent curiosity* is thereby born of *Free Will*, and all on an abrupt and sudden impulse, you decide and choose the latter.

...which is the reason you are in this world,
reading this book!
You chose adventure and independence.

As if in the blink of an eye in timeless infinity, a whisk of impulsive intention grabs your focus and enraptures you in a blanket of desire, "Explore, enjoy, experience!" The innocent deed is done; the entangle triggers—a wish is fulfilled—your *Karmic chain* of events sprouts, and your journeys through the material universes begin.

...and so, also begin
the shackles of material existence

Although from the utmost innocence of curiosity, in a test-drive of your free will, your primordial desires began to unfold. A new reality manifests to let you experience and take the reins of a totally separate individualized voyage. The dream unfolds. The spinning of the web, a *karmic entanglement*, is set into motion. With this new focus, a dream-like existence begins to roll out and unravel before you—launching your journey outward bound— away from *Vaikuntha*.

The first material
elemental substance (ahankara)

And so, within a seemingly pure innocent intention from a living wisp of consciousness (that's you), surfaces the first foreign, material substance to the soul—*ahankara*[52], the material ego—*the first of the 24 elemental substances to encapsulate you.*

You now begin to spiral and cascade your sublime existence down through transformations of consciousness and *free-will* into manifest-matter and material-situations. Now you are inside one of millions of universes—exploring birth and death in all species and moving from *chapter of life* to *chapter of life.*

> You have landed in the material world!
> Your Karma has begun.

This new situation is, in fact, a *dream*, according to the ancient 'Texts.' You, *the eternal living entity*, have embarked on a daydream of millions of lifetimes, which naturally equal to zero in the eyes of infinity. There existed no time, or space, where you previously hovered in playful flight above all existence. You were an infinite spark of life in an endless expanse called *Vaikuntha*—"that realm free of exhaustion and free of *karma*." Now, however, you are part of and intertwined with this new temporary *other* existence, a new destination, an alone experience.

---

52   *Sanskrit word for false ego meaning literally 'I-me-mine'.*

Your exploration and individuality from the collective have manifest at your own command and *free-will*. The entanglement of your new chain of events, however, is not as easily undone.

The Puranas explain. You chose this journey of independence and curiosity away from Vaikuntha by *desire alone*. In some distant future, with *pure desire alone again*, you may return to your flight home on your path back to Vaikuntha.

However, the entanglement of *action and reaction* that was set into motion combined with the dream state of forgetfulness from the false material ego, *ahankara*, renders the re-awakening virtually impossible. And, as Shakespeare would say, *"Ay!! There's the rub!"* The difficulty of that *pure desire* required for the return journey is described in the *Bhagavad Gita,* as Krishna tells his friend Arjuna:

> *Out of many thousands among 'mankind'*
> *[barely] someone will endeavor for perfection,*
> *'and as well of those endeavoring', ***
> *and, even out of those who find perfection,*
> *hardly one will become conscious of Me in fact."*

~*Bhagavad Gita Purana 7:3* * यततामपि [53]

---

[53] For the Sanskrit scholars, this third value in the equation is often missing from some translations (यततामपि), adding an additional variable to the level of difficulty.

The return journey, therefore, is much more complicated. This trip back will require freeing oneself from the conditioning and misconceptions from thousands of millions of lives *in as many bodies and species*—across '*who knows how many*' universes.

A sincere dedication is subsequently required to return home, one that you sow and nurture from lifetime to lifetime with *perseverance in consciousness*—the *adhikar (spiritual wherewithal)*, as mentioned earlier—with a little help:

> "...by the process of constantly endeavoring for
> perfection with the help of
> scriptural evidence, theistic conduct,
> and perseverance in practice."
>
> *~Brahma Samhita Purana 5:59*

In the chapter that follows, we will discuss in greater detail your *dharma—your infinite nature*. This infinite nature, as you travel, remains untouched by any temporality. However, the new transformation, *ahankara—false ego*—is a temporary material element with its own temporary acquired *dharma*. You no longer relate to or are able to see your eternal *dharma* and infinite nature. The false ego (*ahankara*) blinds you. In fact, a form of amnesia begins to set in. The dream slumber continues.

This new combination produces a constant state of incompetence and incompleteness, always awaiting resolve and next steps—a duality. A cascade chain of action, response, and result (*thesis, antithesis, synthesis*[54]) unfurl ever-changing in perpetuity—*karma*. Thus, the living being (*jiva-atman*) becomes engulfed with the new temporary nature. This transitory *dharma* is called in Sanskrit *naimitika-dharma,* acquired *dharma.*

And so, the incoming journey continues until we complete our cascading entrance into the material world. From being pure consciousness to sculpting and building your various bodies and adventures, your new *Relationship* with matter guides you.

You first summoned the *24 elemental substances* of material creation—described in the Puranas—to surround you and encapsulate you into your very own energy field, which some call the *Aura.* This distinct field is described in the Puranas as the "subtle body."

☞ *Note: These 24 elemental substances are not related to the Periodic Table of Elements from classical physics. They are instead consciousness-based plenums, manifestations of ever-so-subtle to gross matter transformations, of that same all-pervading consciousness.*

----

54 Hegel's "Dialectical Triad" will be discussed – http://bit.ly/DialecticTriad Wikipedia

The Sanskrit word *sambandha*[55], meaning the Cosmic *Relationship* between all objects and things (material and spiritual), *includes* our *Relationship* with both the Complete Whole, these 24 substances, and our *Aura—our subtle body—*and *are* the driving forces of what takes place next.

This bridge is when our entrance into matter transforms the early stages of *Cascading Material Development.* A cosmic emergence takes place *in* mutual-relationship *and due to* mutual-relationship, with all things (both material and spiritual), as we continue to explore the Puranas.

But first, we need to bring deep into our meditation the word and the meaning of *dharma*—another very misunderstood *Relationship*.

Dharma, sambandha, karma...

Three fibers of the thread
that weaves the fabrics of reality;
avarice, survival, fear, tribalism,
racism, cults, offense & defense.

---

55   *Sambandha: Relationship between objects and things – https://bit.ly/Sambandha*
      WikiPedia Sambandha

Ganesh - Considered a "Demigod" is the *shakti* of success and remover of obstacles.
Relationship with this *shakti* helps one focus on success
and detachment from toxic situations and people.
(*Photo-Illustration by the author, James Ordonez, from personal photos*)

Chapter 4

Dharma First? Well, OK then!

The unique
original nature
of an object, particle or being

" *Dharma means essential nature. As heat is essential to fire, the dharma of all creation is to love. We may not all love the same thing, but love we must. Ultimately, the dharma of all beings is to love God.*"

~ Swami Bhaktivedanta

You've probably heard of "The Secret,"[56] the 2006 documentary and subsequent book about the "Law of Attraction" and taking charge of your destiny. And, of course, you've heard of Yoga—who hasn't?

What you probably haven't been told (inadvertently) that both "The Secret" and the Yoga conception of existence come from very similar origins and share many of the same ancient philo-

---

56   *The Secret Documentary - http://bit.ly/The-Secret-Documentary*
      Wikipedia The Secret (2006 film)

sophical concepts. This common ground is the vernacular that we will explore here to give light and clarity to both *dharma* and *karma* concepts. This new lens will be our fresh perspective on *action* (*karma*) as well as the building blocks of the universe, from the aforementioned Ancient Texts.

As introduced, the essential source of reference for a discussion on this work are the *Vedas*, the *Puranas*, and the *Upanishads* from ancient India. These ancient writings have driven various schools of thought throughout the ages, like *Buddhism*, Taoism, *Zen*, *Hinduism*, modern *New Agers'* progressive thinking[57], and even discussed in Deepak Chopra's book "Quantum Healing."[58]

Interestingly, as you may know, both the book and movie "The Secret" and the portion of the Puranas that deal with existence and the cosmos receive significant corroboration and validation from modern Quantum Physics. Bits and pieces of this corroboration have been scattered all over the Internet in featured documentaries since the late 1990s. Both blockbuster movie documentaries, "The Secret" and its predecessor, "What the *[Bleep]* Do

---

57   New Age Movement – http://bit.ly/New-Age-Movement
      WikiPedia New Age
58   Deepak Chopra – A recommended Read – http://bit.ly/Quantum_Healing
      Quantum Healing

We Know,"[59] are among these. However, what has been critically missing is the complete primordial perspective for these presentations, as explained initially by the Puranas. The word Purana means *ancient*, and the Puranas are indeed recognized as the most ancient of all writings by renowned scholars.

The Puranas and some modern Quantum scientists propose that everything comes from an all-pervading, *eternal universal consciousness* that is both collective yet individualized. This concept is sometimes replaced or labeled as *"decoherence"* by mainstream scientists in Quantum Physics. In any case, a sort of animation principle behind the mechanics of all things persists in both ancient and modern science.

Professor Richard Conn Henry, Ph.D.., writes, *"Physicists shy from the truth because the truth is so alien to everyday physics. A common way to evade the Mental Universe is to invoke 'decoherence'* [60]*— the notion that 'the physical environment' is sufficient to create reality, independent of the human mind. Yet the idea that any irreversible act of amplification is necessary to collapse the wave function is known to be wrong: In*

---

59   *What the Bleep Do We Know!? – http://bit.ly/What_The_Bleep*
      "It's time to get wise"
60   *The Mental Universe - Richard Conn Henry – http://bit.ly/The-Mental-Universe*
      The only reality is mind and observations, but observations are not of things.

'Renninger-type' experiments,[61] the wave function is collapsed simply by your human mind seeing nothing. The Universe is entirely mental. The world is quantum mechanical: we must learn to perceive it as such. The Universe is immaterial—mental and spiritual." – *Professor Richard Conn Henry, Ph.D.*

The significant difference between Professor Henry's statement and the Puranas' conclusion is his exclusivity to only the "*Human*" *mind*. Instead, the Puranas indicate this "*Mental Universe*" involvement as a universal "*mind substance,*" an energy pervading all existence. As in Buddhism, our ancient texts refer to *Mind* as a ubiquitous *field* of consciousness strewn everywhere regardless of corporeal or specie affiliation—*all life*.

In any case, "*Mind*" continues to be regarded as "a substance or field" across several schools-of-thought and science. This all-pervading field of primordial consciousness, or '*decoherence*,' is believed by some quantum scientists—*and the Puranas*— to be the substantive weave across the fabric of reality. In a nutshell, consciousness is said to be the basic foundational and causal principle of everything in all multidimensional creations, according to various sources.

----

61   *Renninger-type Experiments – http://bit.ly/Renninger-type-experiments*
      WikiPedia

In fact, before the discovery of the *"Higgs Boson* quantum particle"* in 2012, it was theorized by some Quantum Physicists that the *"Higgs Field"* was the ultimate source of the *space-time continuum*. The hypothesis was that this Higgs field was the underlying fabric of all matter—coining it *"The God Particle."*[62]

However, after this Higgs Field discovery, newer scientific theories surfaced, revealing additional evidence of a more interactive and complex foundational probability of the fabric of reality. Subsequently, several physicists theorized an infinity of *"multi-verses"* and infinite multi-dimensions beyond our collective empirical bandwidth of comprehension.[63]

> "After separating the different universes,
> the universal form of Maha Vishnu, from the Causal
> Ocean, entered into each of the universes, wishing to
> repose created the transcendental waters
> [the Garbhodaka field]."
>
> ~ *Srimad Bhagavatam Purana 2:10:10*

Interestingly enough, with all this—despite *decoherence* and the aftermath of the Higgs Boson discovery—Modern Quantum Physics seems to align theory after theory with the ancient

---

62   *The God Particle – http://bit.ly/The-God-Particle*
      "If the Universe Is the Answer, What Is the Question? is a 1993 popular science book
      by Nobel Prize-winning physicist Leon M. Lederman and science writer Dick Teresi".
63   *Quantum Worlds and the Emergence of Spacetime – http://bit.ly/Infinite-Multiverses*
      Deeply Hidden

Puranas—describing a similar eternal reality beyond the subatomic as described in the *Vedic Multiverse of the Puranas*.[64]

The understanding of *dharma* and *karma* lies within and between these two foundations of thought—the evolution of scientific observation and the Puranas' ancient science.

Both old and new sciences describe the observable *Relationships* and reciprocities between all things. That is *sambandha*, the Puranic version of a theory of relativity. Additionally, the word *karma* only means "action," as opposed to popular misconceptions that it means a law of nature. Let's be clear; the word *karma* does not mean "Cause and effect," as many pundits often suggest.

The contextual philosophical inference of *karma* indicates that every action has an owner. For every action, there is a desire that initially manifested as thought, which in turn propelled an intention. As such, that intention was *allowed* (or facilitated) to manifest into action. The term *"allowed"* is the operative word here, as the unleashing of a reactionary chain of events.

The living being, whose original intention created the action, is the owner of the action, or the *karma*, and is linked in *Relationship*

----

64   *Dark Energy Anyone? – http://bit.ly/Swami-Tripurari*
      <u>*The Harmonist*</u>

with the action and all its cascading circumstances and intermingling reactions. That, in a nutshell, is the law of *karma*—the ownership of the action.

Dharma, on the other hand, is the inherent nature of an object, action, or intention. It is also the very nature of the owner of an idea. The science of the Puranas explains that *"an object [particle or field] is called a vastu, and its 'eternal nature' is known as its nitya-dharma (eternal nature)."* Guru Thakur Bhaktivinode explains that *"Dharma nature arises from the elementary structure of an object and remains inherent in that structure as an eternal concomitant factor."*

The Puranas explain that this *original nature* of a given object (particle or being) becomes altered or distorted when a change occurs within it. That will happen either by force of circumstance or due to contact with other objects and their *inherent nature(s)*. Guru Thakur Bhaktivinode explains, *"Gradually, with the passage of time, this distorted nature becomes fixed, and appears to be permanent—as if it were the eternal nature of that object."* This new distorted nature is not the true nature. It is called *nisarga*, or that nature that is acquired through long-term association. This *nisarga* occupies the place of the factual nature and becomes falsely identified as the true *dharma*.

☞ *Our eternal dharma is to "live eternally;" our acquired dharma is the temporary illusion of birth and death, which naturally manifests as fear, avarice, survival, defense, tribalism, racism, cults, offense, & war.*

Guru Bhaktivinode elucidated with a relative Earthly example. *"Water is an object and its true [Earthly] nature is liquidity. When water solidifies, due to certain circumstances, and becomes ice (or vapor), the acquired nature takes the place of its inherent nature. In reality, this acquired nature is not eternal; rather, it is occasional or temporary. It arises because of some cause, and when that cause is no longer effective, this acquired nature vanishes automatically. However, the true Nitya-dharma is eternal. It may become distorted, but it still remains inseparably connected to its object, and the original nature will certainly become evident again when the proper time and circumstances arise."* – Jaiva Dharma (Chapter One)

In other words, this true *dharma* of an object, particle, or being is its particular and inherent eternal function. At the same time, its acquired nature is its occasional function and is temporary. According to the Puranas, by accepting and embracing this knowledge of *dharma*, one can know the difference between eternal and occasional *function* and *nature* of a *creation*.

Guru Bhaktivinode continues, *"Those who lack this knowledge consider acquired nature to be true nature, and they consequently mistake the temporary dharma for eternal dharma."*

Simply described, the eternal nature—*nitya dharma*—of the *jiva-atman living-being* is "eternal truth, infinite consciousness, and endless blissful peace." This precedent, loosely, is the meaning of the popular Sanskrit phrase *"sat-chit-ananda,"* the eternal *dharma* of everything. In fact, the Puranas (and some forms of Buddhism and Shamanism) push us to explore that since everything comes from consciousness, all things indeed are, in one manner or another, a *"person"* traveling the cosmos as you are, be they human, animal, plant life, planets, mountains, rocks, clouds, or even objects and forces.[65] All, coming from and composed of *shakti* (*energy*), share in passing *dharma transformations* from the eternal function of pure consciousness. They acquire temporary functions and attributes for various material existences and sojourns—only to one day return to pure infinite consciousness.

---

65   *Shamanism - http://bit.ly/Everything-is-a-person*
      <u>*Shamanic Healing*</u>

" Since everything is a unit of consciousness, everything has personal existence. Everything is a person. Before we go to the material conception, we must pass through the personal conception or aspect of that thing."[66]

~Swami B.R. Sridhar - Subjective Evolution of Consciousness[67]

For those interested in the theosophical theories of the Puranas, if you read only one book on the subject, that is the one, "Subjective Evolution of Consciousness"[68]

~by Swami B.R. Sridhar.

...and as if by Coincidence?

In the Quantum world, *eternal dharma* vs. *acquired dharma* shares some relation to String Theory. In Quantum Science, anything divided by zero equals infinity, and dividing anything by infinity equals zero. But, they have to cheat; zero and infinity don't especially *exist or have consequence* within the Quanta Mathematics of the material world when trying to measure

---

66   Consciousness and the Self X – http://bit.ly/Self-X_Consciousness
      Shadow Consciousness and Rahu

67   Swami B.R. Sridhar – http://bit.ly/Subjective_Evolution_of_Consciousness
      Subjective Evolution of Consciousness

68   Swami B.R. Sridhar – http://bit.ly/Subjective_Evolution_of_Consciousness
      Book – "Subjective Evolution of Consciousness"

"zero-point' energy, e.g., electrons.[69] On the other hand, in our logic extrapolated from the Puranas, both infinity and zero only truly exist in pure consciousness, before *ahankara,* before the *24 elemental substances* engulf consciousness. Before *Maya* (the word literally meaning *"the course [ya] of measurable [ma] reality"*),[70] therefore, before *ahankara*, there are no *measures, distances, or sizes.* Everything is zero and infinity simultaneously. Back at our "fork on the road," upon entering the material expanse, *"measurement"* begins. At this point, the *eternal "sat-chit-ananda" dharma* becomes superimposed over by the newly *acquired material, temporary dharma* from outside forces introducing the *24 material elemental substances.* Therefore, the Puranic solution is to return to *zero infinity*— *"home"*—one day.

Coincidence??

On a related note to *"acquired dharma,"* String Theory solves a central Modern Science paradox with zero and infinity in reconciling General Relativity and Quantum Mechanics. In a very similar perspective as *"eternal dharma,"* strings being identical by nature acquire temporary and occasional traits when associated with other outside forces:

---

69   *Foundation for Mind-Being Research - http://bit.ly/Zero-Infinity*
      The Mysterious Zero / Infinity
70   *James Robinson Cooper – http://bit.ly/Maya-Meaning*
      Facebook - The Language of the Gods

> *The leading approach to unifying quantum theory and general relativity is string theory. In string theory each elemental particle is composed of a single string and all strings are identical. The "stuff" of all matter and all forces is the same. Differences between the particles arise because their respective strings undergo different resonant vibrational patterns – giving them unique fingerprints. Hence, what appear to be different elementary particles are actually different notes on a fundamental string."*
> *~ William C. Gough , Mar 2002*
> *Foundation for Mind-Being Research[71]*

To understand the Magic of *karma* and the simplicity of all our origins, one must understand this fundamental *dharma* and be able to identify *Relationships (sambandha)* between *temporality* and *permanence*.

As this *dharma* concept trickles into realization, by meditation, the basic understanding of *eternal dharma* and temporary *acquired dharma* starts to show us Universal Relationships and reactions thereof—*karma*.

· · ·

In regards to the tribal herding instinct of cults and twisted lies for racist agendas, understanding *dharma* plays a significant

----

71   *Foundation for Mind-Being Research - http://bit.ly/Zero-Infinity*
      The Mysterious Zero / Infinity

role. The Puranas continue to explain that the *eternal dharma* nature of all things is *"eternal, infinite, boundless consciousness.* Therefore everything is *dialectical* by its own *eternal dharma nature.* This concept needs to seep deeply into our holistic understanding of all things. This is that *"upside-down, inside-out, and backward"* lens we discussed earlier. It isn't easy with our Western upbringing and genetic memories[72] to see and understand until this realization takes place—*"everything, everything, everything"* is dialectical and eternal consciousness by nature. The Puranas press us. Until we truly accept this wholeheartedly, we will not understand the acquired temporary—yet—inherent nature of material life, people, agendas, politics, cults, and hatred.

Not only are our intentions, actions, constructions, and matter composed from *dialectical consciousness,* but so are situations, tribes, cultures, movements, governments, beliefs, and—yes—conspiracy theories and political and religious agendas. They are born from faulty sense perception, unconscious bias, and competitive "Motivated Reasoning."[73]

As we stitch all this together from the Puranic perspective of

----

72 *Genetic Memories – http://bit.ly/Genetic-Memories -*
73 *Motivated Reasoning – http://bit.ly/Motivated-Reasoning*
     Psychology Today

universal dialectical consciousness, some[74] will begin to truly see a semblance of this truth from the *Ancient Texts*.

Next, as we continue to study the Puranic *sambandha*, we may see the relationships of mindsets, mass-mind, cults, cultures, and movements with both lenses, *eternal dharma,* and *temporary acquired dharma*. And there is a silver lining.

Is there a silver lining? We may ask ourselves. "What would happen if we were to switch our educational systems (Kindergartens through College) away from separatist/racial world religions—*"that all others are evil or lost and are going to some hell or another?"* Might that be a start? Imagine an educational foundation of scientific-based Situational Ethics and simple non-sectarian values and principles of Karmic Cause and Effect? Likely soon? Doubtful! Possible, of course, but religions and cults need to be put in their place—away from education—maybe in a distant future.

Yes, putting completely aside the Puranic "Theosophy," as separate from the Puranic scientific karmic theory!

---

74   *We remind the reader about "adhikar," the voluntary indulgence in "spiritual wherewithal."*
      Adhikar – http://bit.ly/Adhikar-Qualification
      Veda.WikiDot – Adhikar – In our context, a spiritual qualification and wherewithal that adorns and guides the individual drop of consciousness through the journey

If theories in theosophy would be taught as "theories" alongside scientific theories, perhaps a kinder and gentler humankind might emerge amidst the chaos of our past genetic fields and acquired *dharmas*. While unlikely for the masses, the individual may consider this for *self-care, family care, and community care*. And for discussion sharing in relationship with those around them— might be a start. Isn't it time to get real in each of our lives!!

Dharma (In a nutshell),
...leads "Relationship" in universal scale
(sambandha)

☞ *THE CHAPTER THAT FOLLOWS*
is long, and by design tedious and repetitive
to drill and drive some challenging subtle perspectives
with which a lot of us are not yet accustomed.
STRAP IN TIGHT!

...there's gonna be a quiz...

Sarasvati - The shakti of learning, music, harmony, and rhythm.
May be a meditation for artists and creative individuals
(Photo-Illustration by the author, James Ordonez, from personal photos)

Chapter 5

The Magic of Universal Relationship

Enter Sambandha![75]

66

The mind (any mind) is a substance, and it is possible that it exists without the body (any body), and the body (any body) is a substance, and it is possible that it exists without the mind (any mind)."

~ *Descartes' Conception of Substance*

I t is vital to note that many Buddhist interpretations of these ancient texts speak of *Mind*, not as "individual minds." Instead, the collective all-pervasive *Mind* is strewn infinitely across space and time. In one manner of speaking, *Mind* is a foundational essential material *substance* of life and all living beings within the material universes. According to the Puranas, and apparently

---

75   *Sambandha – http://bit.ly/Sambandha*
      Universal Relationship between cosmic energies (shaktis),
      living beings, and the Cause of All Causes.

Descartes and others, *Mind* is an all-pervading substance and is listed among the *24 Puranic elemental substances* of creation.

As mentioned, the Sanskrit word *Sambandha* means 'Relationship' between all objects and all things (material and spiritual). This Puranic theory of "relationship" reflects Einstein's classical physics theory, defined as "Einstein's theory that time and space are not absolute."[76] *Sambandha* also is the Puranic version of a theory of relativity, explaining how things only have value *"in relation to other things,"* which is the opposite of absolutism.

It's no secret that Einstein, Descartes, Jung, Hawkins, and many other great minds in our scientific community studied the Vedas and Puranas. They would often refer to the Puranas, and there is sufficient evidence all over the Internet Library that they very often borrowed modern theories from Ancient Texts. Inspired or copied, could they all be referring to the same ancient theory of relativity, *sambandha*?

The story we are building has several characters in development—the soul (individual consciousness), free-will, choice, and *ahankara* (the ego concept of I-me-mine), among others. The latter (*ahankara*) leads the way as the first of the *24 Puranic*

----

76   *Everything is "Relative" to relationship and perception – http://bit.ly/Theory-of-Relativity* Dictionary.com

elemental substances from consciousness composing material creation.

These *24 substances* are the *elements* that house the ecosystem for *consciousness* to engage with and revel in the material universes. Relationships between these *24 substances* vary according to their *eternal dharma* and such *acquired dharma* imposed upon by association with matter.

Grasping this science of *sambandha* (Relationship) is of paramount significance for understanding anything from these ancient texts. What sadly happens in most (if not all) of the groups and cults who study the Puranas and Vedas is precisely that, not having prioritized this Puranic Science of '*Relationship.*' Jumping over, they miss the point entirely.

Guru Bhaktivinode obsessed on this detail, "*Without a solid grasp of Sambandha all one's bhajan (attempts†) remain on the neophyte level.*"

† Bhajan means "offering reverence" and loosely, 'prayers or meditations'.

This solid understanding of *sambandha* Relationships is the actual key to the locking mechanism. Rather than deciphering the scientific messages inscribed in the translations, cults and religious groups instead get literally stuck on sentimental interpretations. We have seen this phenomenon very much resembling a chicken stuck on a chalk line. They get caught up on the literal language translations—in the illustrative narratives—with Earthly perceptions *(points of reference)* of *human* love, beauty, form, color, emotions, and even superficial ethical or cultural anecdotes. *Again*, the cults remain totally stuck, bypassing the main point.

Although in this book we focused on the science of *sambandha,* some references to theism are essential for balance with all the writings and assertions among the world's Puranic scholars. After all, the Ancient *Quantum-like Science* is ultimately based on the Hindu divinity of creation and its *Relationship* with the Complete Whole—God.

Arguably, the Puranas propose a profound notion, God's ubiquity is powerful enough to exist as manifest story form, or *"Lila"* *(pastime parables)*. Some scholars will argue that all things being equal, the explicit face-value of the mythology is 'in and of its own' a worshipable deity of sorts. As much as a statue or object can represent a deity on an altar, a recital as well may represent the divine subject. This premise does make some sense by the profound Puranic explanation of *universal-relationship* in *sambandha*. They, *God*, are, after all, *omnipotent* and *omnipresent* in everything. However, the point is that some cults' subjective (mundane) perceptions through "free-will" are so tainted (with politics, greed, recognition) that a much less than perfect human nature remains insidiously lurking abundantly within some groups. Some tribes become riddled with politics, competition, money, fighting, crime, sectarianism, and messianic delusions. They predominantly remain obsessing that they "...are the only ones *saved*!" And again, the cult remains completely stuck in sentiment, bypassing the main point altogether. Shackled in the illusion of narcissism becomes their temporary acquired *dharma*.

At that juncture, the group becomes a "Religion" or "Cult" ruled by misplaced dogmas or idiosyncratic cultural peculiarities and taboos from the past—rather than receiving the intended under-standing of the Puranas or scriptures.

☞ *After all, religion, with all its war, judgment, history, dogma, sex, gossip, money, power, self-aggrandizement, and taboos, are far more exciting and materially stimulating than drab observations on the science of 'Relationships' between consciousness-based substances.*

More precisely, the idea of *Relationships* between *substances (shaktis)* based on consciousness is too foreign and mind-bog-gling to our materially conditioned faculties. Strict discipline is required to carry and maintain the perception moment to moment, what to speak of day-in and day-out. No doubt, this need for discipline by sincere monks over time contributed to the birth of the many diverse schools of thought branching out from these ancient texts—Taoism, Zen, Buddhism, Bhakti, Yoga, and many factions among each modality.

Therefore, for our purposes of discovering the Magic of Karma, we focus primarily on the *sambandha* science of relationships between the *24 material substances* from *consciousness* listed in the Puranas. As mentioned, these elements are not related to the Periodic Table. Accordingly, as per Descartes' "Conception of Substance," the Puranas, and other sources, we need to view these transformations of consciousness as *elemental substances* or *plenums*—as in vibrational energetic *fields*.

Here is where a religious group may argue that "impersonalism" is the Devil that destroys one's spirituality—our *"Poster-Child Cult,"* of course! Indeed, this is born from the *Hermeneutic Religious Literalism* we discussed earlier on.

Nevertheless, with this mindfulness meditation of *sambandha* and *elemental substances*, we continue reminding ourselves of our very first material Relationship. That is when the *jiva-atman* living being first meets the material ego, *ahankara*, born as merely a wish for independence, back at *"the fork on the road."*

...and so, ahankara engulfed the Soul...

It is there that a *wisp of consciousness* enters the material world. At that cosmic moment, the *24-substances* begin to cascade and manifest around the living-being. These altogether form the subjective field of activity—*the aura*. We will not attempt to explain all *24-cascades* in detail yet in this edition, as that will create voluminous discussion and very speculative argument. We would have to dedicate an entire chapter to exploring each element—perhaps in a future edition. For now, the first few will suffice to give both a meditative and comprehensive study.

Tighten your seatbelt!
We go even deeper down this rabbit hole

Our *material existence* began at the *"fork in the road"* where the living entities chose the path of exploration. An important focus for our contemplation is to consciously remember that the choice was indeed ours, as indicated in the Puranas. Typically, when we lose our way, we must trace our steps back to the beginning of a journey. Therefore, as part of our ongoing meditation, we continue to revisit the *"fork in the road,"* where we turned away. It is there that the first foreign substance to the soul, the *ahankara* bubble—the *"I-me-mine"* ego-substance—engulfed you in the ocean of consciousness.

We sometimes envision this material ego as a suds bubble rising above the ocean, floating away, moving along with a breeze to an undetermined destination. And so begins the first of *24 cascading* transformations of the *jiva-atma—the Ahankara* material ego engulfing the soul.

At this primordial fork, the very first *material substance* manifested and is now in contact with the *jiva-atman* for the first time. There is little chance of turning back since the fulfillment of the desire has already begun. A pure unadulterated spark of life is now entangled within a *transformation* of consciousness, changing your perception from *"I as part of the whole"* to the *"I-me-mine ego"*—the beginning of the journey—*a Karmic Entanglement*. The Relationship between the self—*jiva-atman*— and the false material ego has begun.

☞ *Arguably, since this Relationship is taking place beyond space and time—the jiva-atman and the first material substance—we might envision or compare this phenomenon to the theory of a Quantum Entanglement: "...a quantum mechanical phenomenon in which the quantum states of two or more objects have to be described with reference to each other, even though the individual objects may be spatially separated."[77]*

The well-known Hegel Dialectical Triad—***thesis, antithesis, synthesis***—plays a significant role in what appears like coinci-

---

77   *Science Daily - Quantum Entanglement – http://bit.ly/About_Quantum_Entanglement*
      Quantum Entanglement

dence.[78] Most of us agree, there are no coincidences. First, since everything is consciousness, then the various cascading *transformations* of consciousness resemble a systematic *dialectical* back and forth set of responses—*a consciousness cause and effect.*

Hegel's Triad suggests that for every idea, proposal, or wish—*the thesis*—an immediate concomitant challenge opposing becomes manifest—*the antithesis*. Likewise, that challenge elicits a solution from the combination of the two. Hence, *the synthesis* arises and becomes manifest—*a thesis anew*. In turn, once again, as a new *thesis*, it provokes an *antithesis*, and so on—ad infinitum.

> Such are the simple mechanics of creation,
> dialectical from consciousness,
> proposed by the Puranas

In a nutshell, this initial *recursive cascade* of elemental substances is the primary primordial unfolding chain of events set into motion as you, the living being, enters and travel the material worlds' expanses.

Another "*coincidence*" is that all this recursive chain of events

---

78   *The Dialectic Triad – http://bit.ly/DialecticTriad*
      <u>Wikipedia</u> *"The concept of dialectics was given new life by Georg Wilhelm Friedrich Hegel (following Johann Gottlieb Fichte), whose dialectically synthetic model of nature and of history made it, as it were, a fundamental aspect of the nature of reality (instead of regarding the contradictions into which dialectics leads as a sign of the sterility of the dialectical method, as Immanuel Kant tended to do in his Critique of Pure Reason)."*

also follows the Golden Rule[79] of the Fibonacci mathematical sequence, sometimes associated with the Golden Ratio and The Golden Spiral. The basic premise found by Fibonacci is a ratio in nature, where formations also closely follow a recursive cascade.

The Fibonacci rule finds that the spiral of petals on a daisy, the development of galaxies, and even the reproduction or proliferation of species follow the Golden Spiral order. The rule is the mathematical ratio of "1.618 to 1"[80], where the natural growth of things cascades in a spiral ratio of the present and past states, added together to form the future next state.

As with prime numbers, whose rule is that each number can only be divided evenly by itself or by the number one, Fibonacci numbers can only be the sum of the two previous whole numbers, i.e., *1, 2, 3, 5, 8, 13, 21, 34, 55, 89,* etc. Fibonacci found this ratio in natural transformations of growth that cascade by adding the preceding value to the present value, which can only be the sum of the two previous states. Even today, with modern observation, this rule holds true. This Golden Ratio can be seen in the spiraling development of organic progressions in growth, among all things.

79   *The Golden Rule known as the Fibonacci Sequence - http://bit.ly/TheGoldenNumber*
      GoldenNumber.net
80   *Interestingly the Golden Rule ratio of 1.618 was originally discovered*
      in ancient India by the early mathematician Bhaskar - http://bit.ly/The-Mainstream-of-
      Mathematics
      The Mainstream Of Mathematics by Edna Kramer 1951

☞ *Please note that both Hegel's Dialectical Triad—"thesis, antithesis, synthesis"—as well as the above Golden Spiral follow the same simple equation of adding the present value to the previous value to create the new value. Coincidence?*

To be accurate, the origins of these two observations had earlier founders who were not given the proper credit. Hegel's Dialectic Triad was first discussed by Johann Gottlieb Fichte and before Emmanuel Kant in his 'Critique of Pure Reason.'[81][82] Likewise, Fibonacci's Golden Spiral cascade was first discussed around 600 B.C. by a Hindu mathematician named Bhaskar.[83] Therefore moving forward, we may reference these as *The Dialectic Triad* and *The Golden Spiral.*

In any case, these two well-known observations in nature are quint-essential in understanding the Puranic cascading development and journey of the *jiva-atma*—the living being. The cascade becomes the *impulse* as the *soul* enters the material expanse, develops material faculties, and journeys throughout from body to body,

---

81   Stanford - http://bit.ly/Kant-Critique-of-Metaphysics
82   Wikipedia – http://bit.ly/Critique-of-Pure-Reason—Kant

83   Interestingly the Golden Rule ratio of 1.618 was originally discovered
      in ancient India by the early mathematician Bhaskar - http://bit.ly/The-Mainstream-of-
      Mathematics
      The Mainstream Of Mathematics by Edna Kramer 1951

transmigrating across multiple universes. This is a *dialectical journey* of cause and effect—*Karma,* according to the Puranas.

Recapping

Back at the *fork*, at the exact primal moment when the soul—*jiva-atman*—traveling among the flock within the celestial effulgence, submits to *her* desire for the *free-will* of exploring independence—at that juncture, *her* material existence is set into motion. That first material desire, *ahankara (false ego), the first of the 24 elemental substances*—manifests around the soul.

☞ *Interestingly, some ask why Ancient Texts speak of the soul in the feminine gender, "she" and "her." Certain' religious cults will whisper that it's because the jiva-atman, being single individual sparks from the whole, are therefore subservient — [bleeping] old world gender bias patriarchal reasoning. I've even heard many times "swamis" declare, "God is male because the jiva-atman are subservient." The cosmic joke is that the Puranas teach us holistically when we study Sambandha that God, instead, is primarily not only both simultaneously female and Male by human standards, but that without the female portion, the male aspect cannot even exist. Much more on that later.*

This wisp of consciousness, the soul, a drop in the ocean of eternity, begins to create her very own subtle and material bodies and all the journeys that await.

That bubble we imagined rising above the ocean and going airborne with the breeze has met the first criteria of material development. The soul has been literally engulfed by the first material, foreign substance, the material ego—*ahankara.* Together they drift off into the unknown—the soul and her newfound garment, the covering of *the false ego.*

Suddenly, by innocent curiosity alone, you were encapsulated within and on your own with only one other companion—*your own daydream idea and desire* for exploration and Independence—*embracing you like a bubble as together you drifted on into your unknown journey.*

Just imagine. As that first transformation is complete, *the jiva-atman ponders,* "Now what?" Perplexed, in the dark, reaching and clutching for any answer, resolve, or next step, *"Who am I, and what am I doing here?"* asks the new identity—the new material ego. There is no past; time did not exist. The new ego has no memory. A struggle ensues with the gyrating uproar of this union—a wavering dance of sorts begins, begging for answers and direction.

> ...Jiva-atma plus Ahankara
> Elicit Material Vibration...

Another *elemental substance* spins out from the commotion—material *vibration*—audible as sound. Before the *primordial fork* on the road, we existed as eternal, pure, infinitesimal sparks of life consciousness. We were progressing along in our ocean-like habitat, naturally as part of the *collective consciousness*. From that first combination of the jiva-atman with *ahankara*, the second of the 24 elemental material substances manifests around the soul—*vibration (sound)*. The cascading entanglement with matter continues.

The *Puranas* indicate that this second *substance, vibration*, facilitating *free-will* originally permeates *from* the ocean of consciousness, not as a material vibration but as a *spiritual vibration* originally underlying the *spiritual expanse* where that ocean of consciousness dwells. This vibration is the syllable *Om (AUM)* vibration (*pranava*), which is the meditation chant for many yogis. The *Puranas* explain that this *AUM* spiritual vibration now becomes infused into the material expanse to facilitate the building blocks and materials needed for the *jiva-atman* to create its new home.

Here, that *spiritual sound—AUM—*takes on material qualities and becomes *material vibrations and frequencies—the second material elemental substance*. It is as if—and *is*—the ocean of consciousness willingly and purposefully facilitating the *jiva-at-*

man's journey, with the building materials needed for the sojourn and interactions.

We understand from classical physics and Quantum Mechanics that all spectrum of energies, light, radiation, sound, and even matter are, in fact, wavelengths of vibrations at many different frequencies. While it is unclear how many spectra scientists have identified with a limited number of instruments so far discovered, the Puranas and Quantum Theory indicate the number to be infinite.

From our limited takeaway and understanding, these many spectra of vibrations intermingle between an indeterminate number of dimensions holding the universe(s) together.

☞ *The substance "vibration" [sound] enters the material expanse to facilitate sambandha, all the shaktis (fields), the 24 elemental substances, and the cohesion of the emerging material expanse.* ☜

Our human faculties can only *consciously* observe very few of these frequencies *(vibration)* in light, matter, and sound. Beyond our limited grasp exist ultraviolet, gamma rays, x-rays, ultrasonic, supersonic, infrasonic, etc. And these are only the ones humankind has been able to build sensors to measure. According to the Puranas, these *infinite* vibrations, based on *consciousness*, are the cohesion and substance of all creation.

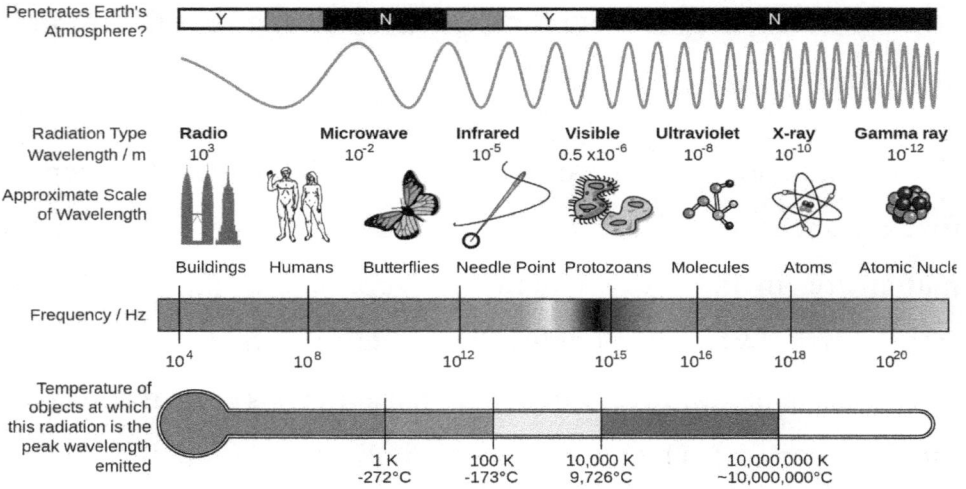

| Penetrates Earth's Atmosphere? | Y | N | Y | N |

Radiation Type	Radio	Microwave	Infrared	Visible	Ultraviolet	X-ray	Gamma ray
Wavelength / m	10^3	10^{-2}	10^{-5}	0.5×10^{-6}	10^{-8}	10^{-10}	10^{-12}
Approximate Scale of Wavelength	Buildings	Humans	Butterflies	Needle Point	Protozoans	Molecules	Atoms	Atomic Nucle

| Frequency / Hz | 10^4 | 10^8 | 10^{12} | 10^{15} | 10^{16} | 10^{18} | 10^{20} |

| Temperature of objects at which this radiation is the peak wavelength emitted | 1 K -272°C | 100 K -173°C | 10,000 K 9,726°C | 10,000,000 K ~10,000,000°C |

NASA Image compares wavelength & frequencies in relation to our visible spectrum
(©NASA. License courtesy of NASA through WikiMedia Commons. See Link) [84]

In the Bhagavad Gita, Krishna tells his friend Arjuna, *"I am the taste of water, the light of the sun and the moon, the sacred syllable Om,* **the sound in ether** *and the ability in man."*—*Bhagavad Gita Purana 7:8*. Each of these *five* are a product of vibration—*a sambandha Relationship*.

We also read that other species often vary in their perception of these spectra of frequencies, with light and sound.[85] This is not to be confused with the myth that humankind only uses a small percentage of their brain for conscious awareness.[86] We simply

---

84   *EM Spectrum Properties – http://bit.ly/WikiMedia-EM-Spectrum*
     https://commons.WikiMedia.org/wiki/File:EM_Spectrum_Properties.svg
     (Color file adapted to gray-scale for printing)
85   *What Can Animals Sense That We Can't? – http://bit.ly/BrainFacts-Animal-Senses*
     BrainFacts.org
86   *Myth Brain Usage – http://bit.ly/Brain-Usage-Facts*
     BrainFacts.org

are not equipped with the faculties of sensory perception to see beyond our built-in limited abilities.

We learn from the *Puranas* that the ability to grasp and understand these topics is a *desired spiritual awareness* beyond sensory perception. Rather than using our limited material senses, this wisdom teaches us to see with that *awareness*—that same spiritual wherewithal we discussed earlier—*adhikar*.

Alan Watts tells us, "*...human consciousness is—at the same time as being a form of awareness, and sensitivity, and understanding—it's also a form of ignorance. The ordinary everyday consciousness that we have, leaves out more than it takes in. And because of this, it leaves out things that are terribly important. It leaves out things that would—if we did know them—allay our anxieties, and fears, and horrors, and if we could extend our awareness, to include those things that we leave out, we would have a deep interior peace."* —The Web of Life (Part 1)

Nevertheless, we all agree *vibration* is the elemental constitution in all things—at different frequencies and spectra. *This vibration—"the sound in ether"*—is the second of the *24 Puranic elemental substances*, which cascade as transformations of consciousness by the *jiva-atman's* entrance into association with the *ahankara*.

This new home, so far composed of the *jiva-atman, ahankara,* and *material [sound] vibration,* is now your developing material foundation for subtle and gross bodies to use as a series of vehicles for the epic journeys that follow.

Swami Bhaktivedanta explains in his book, *Easy Journey to Other Planets,* that it is the living entity who is "the creator" of the material expanse, which she—*the jiva-atman soul*—encapsulates and inhabits throughout the sojourn: ***Easy Journey to Other Planets, Chapter 1***—*"Life is definitely not generated simply by a material reaction like a chemical combination, as many foolish men claim. Material interaction is set in motion by a superior being who creates a favorable circumstance to accommodate the spiritual living force. The superior energy handles matter in an appropriate way—as determined by the free-will of the spiritual being. For example, building materials do not automatically "react" and suddenly assume the shape of a residential house. The living spiritual being handles matter appropriately by his free-will and thus constructs his house. Similarly, matter is the ingredient only, but the spirit is the creator."*

Interestingly enough, an overwhelming majority of Swami Bhaktivedanta's followers read this essential teaching and still completely preclude the *jiva-atman's* involvement having any participation in creation. The fear there, due to Religious Literalism, is that they may begin thinking they are God. Therefore, they elicit the proverbial 'ten-foot-pole' and with willful negligence avoid the Puranic *sambandha* Quantum (like) Science before them, at all costs—*even from their own teacher.*

Still, the *living being* continues to build their new home in *Relationship with* the building materials provided by the Cause of All Causes—the Ocean of Consciousness. Material *vibration* has entered the expanse.

...ahankara plus material vibration elicit Ether...

Yes, ether as in the vacuum of space, also previously known as *Luminiferous Aether.* In Quantum Science, this substance has been upgraded from meaning merely "light-bearing vacuum" to "The Quantum Field," accommodating all vibrations beyond just light.[87] This Puranic elemental substance, *ether,* is the third transformation of your consciousness.

You begin to cause *cascading transformations* around you like bubbles from the most subtle to grosser substances by your mere dialectical eternal nature—*nitya dharma.*

---

87   ResearchGate – http://bit.ly/LuminiferousAether
        Does spacetime possess the properties of a "relativistic aether"?
        Modern discussion on previous deprecation of Luminiferous Aether

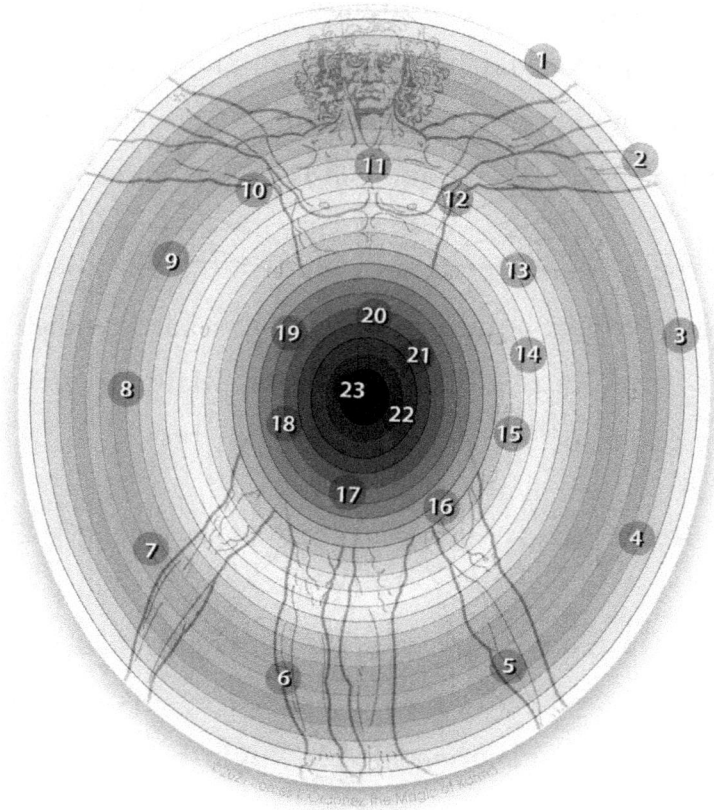

Illustration by the author, James Ordonez.

The Puranas allude that *ether*, the Quantum Field, emerges from consciousness to accommodate vibration (*sound*) in its aggregate relationship with *ahankara* and the *jiva-atman*. At first, this nitya *dharma* is inherent in your existence, and that eternal Relationship is shared within the Ocean of Consciousness—the Cause of All Causes.

Then, by association with those first three primal material elements —*ahankara*, *vibration*, and *ether*— the temporary *dharmas* gradually begin to cover and appear to take the place of your eternal nature.

Taking in a breath...

We need to remember here that these *transformations* and *Relationships* are *ephemeral and phenomenological*, both across the expanse of infinite consciousness and our individual subjective consciousnesses—*altogether simultaneously*. There is no past or present, or back then in time, or my development before yours; there is no semblance of time.

The Puranic *Cosmic Emergence* is a present-moment phenomenon. Beyond time and space, creation is perpetually blossoming ubiquitously and insidiously in all eternity—secretly. It is, after all, according to the Bhagavad Gita, *"the most secret of all secrets."*

What is beyond our grasp and comprehension in these Puranic allusions is just that. This cosmic *emergence* from *consciousness* alone happens simultaneously across all existence beyond time and linear space. That is the meditative sacred moment for the Buddhists and the *syllable Om* from the Vedas and Puranas.

☞ *Can you guess what great minds across eons pondered the "Substance" and effects of Time on our ability to 'see?'*

❝ **The only difference between Future, Present, and Past are merely**

stubborn illusions built into the 'Matrix.'"
~Some Great Minds across time (can you guess who?)

We cannot allow our sensory limitations to become the show-stopper, for that is the very foundation of Maya's illusion—*our inability to see and comprehend beyond our limited and conditioned faculties.* Therefore, we summon by a sincere desire that wherewithal, the *spiritual awareness* to see and carry with us—*adhikar*.

Exhaling

The Puranas continue. These spiritual vibrations and *shaktis* that constitute the totality of everything within the universes give the traveling jiva-atman all she needs for the sojourn. These *24 elemental substances* facilitate all aspects of the journey, at the *free-will* of the independent living being, as the driver.

This *facilitating* spiritual totality of consciousness—the "Complete Whole"—is called in the Puranas *"Bhagaván,"* the source of everything. Synonymously, *Bhagaván* is also the effulgence, the *Ocean of Consciousness,* and the Syllable *Om (AUM)*. The Puranas regard this totality of consciousness as the supra-personal intelligence, the *Summum bonum* of all that is—God, the *Supreme Consciousness*. Still, that is for yet another edition of this book series dealing more profoundly into the Puranas' theosophy.

Moving on, we continue discussing the various transformations from consciousness into matter. The original *Om (AUM)* vibration continues to accompany and facilitate the *jiva-atman* in everything that the soul needs for the journeys and sojourns, which await.

Each cascading transformation is a *substance* or field of energy, each with its own temporary and eternal natures—*dharmas*. Just as with light and sound, each substance vibrates at an inherent unique energy frequency, forming waves of crests and troughs, perpetually vibrating corrections from wavelength to amplitude.

> *And so life is a system of now you see it, now you don't, and these two aspects always go together. For example, sound is not pure sound; it is a rapid alternation of sound and silence, and that is simply the way things are. Only, you must remember that the crest and the trough of a wave are inseparable."*
>
> ~ Alan Watts' *Coincidence of Opposites*

Likewise, these *24 elemental substances'* subsequent vibrations are endlessly setting into motion the cause and effects of each and every manifest energy, creation, or activity—*karma*. That is not only the case with light, sound, and energy but with thought, intention, and action, as well as with all the reactions—*precisely as a wavelength*.

We repeat for our meditation

As in Hegel's 'Dialectic Triad' discussed, the *synthesis* resolved from the first two substances—*jiva-atman* plus *ahankara*—naturally creates another *antithesis* reaction, whose byproduct, in turn, is a new synthesis, the *substance vibration*. And the dance continued; *thesis, antithesis, synthesis*, a dialectical chain and stream of consciousness, looping, adjusting and transforming. *Ether (space)* came next to accommodate the resting place for *vibration (sound)*.

Put another way

Back at your *primordial fork*, the resulting chain of adjustments and corrections between the *jiva-atman* and *ahankara* elicited a new *substance*. There, *material vibration was manifest* to accommodate the previous material *Relationship*. Likewise, to accommodate a foundation for *vibrations, ether* was then invoked—the third *transformation of consciousness*.

It is crucial to continue remembering. When we say '*transformations of consciousness*,' we mean transforming from intentions, desires, perceptions, needs, etc. These are the dialectical progressions from the original living *wisp of consciousness*—now in association with the very foreign elemental substances of matter. For example, brick and mortar are manifest from desires, intentions, plans, research, eventually action—*karma*.

These new associations are temporary material substances manifest from a unique and temporary situation, eliciting responses and corrections. Because this is all *consciousness-based*, each subsequent vibration, or *substance*, is a part of the dialectical stream of consciousness accommodating itself—a lonely monologue of sorts.

It becomes apparent in our meditations; we can never revisit enough particular examples and observations in our reflections. Although not stated in the Puranas, it seems of no coincidence that both the '*Dialectic Triad*' and the '*Fibonacci Sequence*' are of direct consequence. In the Dialectic Triad stages of development, the tension between the first Relationship forms the resolve as the synthesis from the combination of the first two—*concept → reaction → solution*. Likewise, we revisit the Fibonacci Sequence, where *the Golden Ratio Spiral* is continuously reborn from the sum of the previous Relationship—the result of the previous two values.

Similarly, the cascading transformations of consciousness develop from the subtlest of matter to gross matter density. The subsequent intricacies of interactivity that follow employ the same principles and mechanics of the recursive cascade as these two discoveries above—the *Dialectic Triad* and the *Golden Ratio*.

The Puranas explain that the 24 material substances are

consciousness transformed manifestations into *matter*, as 'willed' through these transformations. They begin when the living being meets the *I-me-mine ahankara*—the material false ego.

As all this gets more and more complex and ever so subtle to grasp, we begin to understand why the Puranas and other Ancient 'Texts' were written down in the vernacular of lore and legend—mythologically. Similarly, as we develop this book, we look to overall examples and terminologies to enhance our comprehension by illustrative metaphors.

> "The Tao that can be spoken of
> Is not the Everlasting Tao
>
> The name that can be named
> Is not the Everlasting name
>
> The nameless is the beginning
> of heaven and earth
> ~Tao Te Ching, Lao Zu

The edict from *Taoism* generally explains this level of difficulty in laying everything out and trying to discuss with our temporary and limited faculties that *the truth cannot be spoken or explained*. We, therefore, wrestle with this middle ground in understanding what we can only—waiting for the rest to come internally— through realization from contemplation. The *Tao Te Ching* loosely indicates that this *dialectical* foundation of all things is *phenomenologically* beyond measure or grasp—the magic of *Maya*.

Essentially, as these topics of the *truth* in consciousnesses and their dialectical transformations begin to surmount, our empirical jurisdiction of experience begins to clog up and digress—losing focus and cohesion. This enigma is by design—*Maya*—the grand illusion. Back at the *primordial fork* on the road, when we chose independence over continued immersion and surrender, we elicited support from the 'Cause of All Causes.' We requested a separate, distinct reality and a new expanse of existence that would challenge infinite-reality for a temporary exploratory situation. By conjuring up an illusory, temporary existence, we were, therefore, granted the greatest of all illusory energies—*Maya*—the direct shakti and expansion of the Ocean of Consciousness.† Therefore, only by genuine desire to know could one put it all together and begin to understand through *internal realization*.

(† The illusory shakti Maya, and the other female demi-goddesses, according to the Puranas, are ultimately expansions from the female aspect of Krishna or Vishnu, the "internal potency," also known as Hladini Shakti.)

...

The Cascading Substances From Subtle to Gross

    " *Modern science certainly accepts earth, water, fire,*
    and air as forms of material energy, and ether

*might be so accepted if we were to identify it
as Einstein's curved space-time continuum."*
~ *Physicist Richard L. Thompson, Ph.D. God and Science*

It's important here to continue remembering that these initial primordial substances are literally based on *consciousness* alone. As in a dream, they are elusive and fleeting, as if thoughts and ideas seeking resolve—as in the *cloud of unknowing*[88]—soon to become substantial, situations, and even solid matter. For our purposes, we need to view all this as chronological, i.e., first came the *jiva-atman*, then the choice for independence, then the accommodation of manifest energies. In reality, however, according to the Puranas, existence is happening (emerging) simultaneously in the present moment, beyond time and space.

Professor Richard L. Thompson, Ph.D., continues. "*The Bhagavatam [Purana] (3:26) presents an account of the Sankhya Philosophy, in which the elements of gross matter (air, fire, water, and earth) are described as successive transformations of the* <u>ether</u>. *The sequence of transformation is as follows…*"

"sound ▸ ether ▸ touch ▸ air ▸ form ▸ fire ▸ taste ▸ water ▸ odor ▸ earth
~ *Physicist Richard L. Thompson, Ph.D.
Mysteries of the Sacred Universe*

88   "The Cloud Of Unknowing", an AD 398 'Text' initiated by St. Augustine, often spoken of by Alan Watts and even Leonard Cohen in his poetry wisdom:
http://bit.ly/Cloud-of-Unknowing

Because the cosmic *substances* are transformations of consciousness itself, our new *lens* requires a significant focus adjustment. The elements or substances *(Bhuta)* are in and of themselves *Relationships* between energies *(sambandha)* vibrating at their unique wavelengths and frequencies—starting from very subtle to very gross matter. Their various interactions create the universal cosmos from within our ubiquitous life force, based on the many cosmic *Relationships'* needs and interactions—ultimately, our *collective desire.*

> The Puranas go on to further explain
> the cascade of the elements.

The five gross 'Great substances' *(maha-bhuta)* are *earth, water, fire, air, and ether.* In the Bhagavad Gita, Krishna declares to his friend, "*I am the **taste of water**, the light of the sun and the moon, the syllable Om, the sound in ether and **the ability in man**.*"— *Bhagavad Gita Purana 7:8.* In this verse, the sense of '*taste*' and the '*abilities*' of living beings are understood as necessary creations *(substances / fields).*

These various sensory *abilities* and even the very *development* of sense organs are indeed also recognized in the *Puranas* as elemental substances or elements of creation—among the building blocks.

In the same powerfully subtle manner, the creation of the senses and sense organs is also understood as transformations of consciousness into *cosmic substances, or fields of energy, vibrating as part of the ecosystem.* The creation of these abilities is also *'Life-Force,'* unfolding realities around itself to exist in the material expanse.

The primary five working senses for sentient human beings are *eyes, ears, nose, tongue, and (tactile) skin.* The Puranas' conclusion suggests that for any living being to reside in the Material World, such working attributes must already be included within the Ocean of Consciousness of the complete whole—a mimesis or mirroring—by definition as part of the very nature of consciousness itself.

For these five senses to interact in *Relationship* with the five material Great substances (*maha-bhuta*), certain *qualities* are elicited in the dialectical cascade and manifest among the 24 substances. These are described in the Puranas as the five "*Sense Objects*" and are *(loosely)* as follows.

S U B S T A N C E S				
Gross	**GROSS** substance	**TRAIT** objects of the senses	**ABILITY** receiving sense organs	**ACTIVE** working senses
intelligence	earth	odor	nose	
	water	taste	tongue	
mind	fire	form	eyes	
	air	feel	skin	
ahankara [ego]	ether	sound	ears	
Subtle jiva atma				

All these **substances** are strewn everywhere unmanifest, (pradhana)
until touched by Time, due to the desire of the jiva-atman.
(Illustration by the author, James Ordonez)

Remember, the Puranas describe these plenums as "*qualities*" born of *consciousness (as in temporary dharma or nature).* Therefore, the quality of *earth* becomes *odor [eliciting the sense of smell],* the quality of water becomes *flavor [eliciting the sense of taste],* the quality of fire becomes *form [eliciting the sense of sight],* the quality of air becomes *feeling [eliciting the tactile senses].* The quality of ether becomes *sound [eliciting the sense of hearing].*

As such, the Puranas explain the five "*Sense Organs.*" The *nose* is the organ of odor, the *tongue* is the organ of taste, the *eyes* are the organ of form, the *skin* is the organ of sensation, and the *ears* are the organ of sound. These abilities to *interact* and to *exist* are the *substances* included in the *24 elemental vibrations*—from consciousness.

Each transformation of consciousness, combined with each new *Relationship,* cascades into yet a new 'field' (plenum of consciousness manifest into energy). This order follows the same natural sequences of the *Dialectical Triad* and the *Golden Spiral* discussed earlier. Altogether with the previous combinations of fields, each instance transforms, and consciousness becomes manifest into yet another material energy—forming the subtle body (*Aura*). This process includes the subjective material mind, intelligence, ego, eventually *altogether* the material body.

☞ *Yes, all this is all still extrapolated from the Puranas, behind the illustrative poetic anecdotes enacted in the sages' teachings.*

At first, we may not see the Quantum correlation. Let's consider. Quantum means to measure *(from the Latin for "how much"),* and Maya means "measure" *(in its initial Sanskrit definition).* In Quantum Science, particles are strings, and strings are particles, and both are plenums or "fields," loosely—as an allusion. In the Puranas, *shaktis* are energetic plenums. The *Jiva-Atman* is likened to a photon or pencil of light. The 24 elemental substances are alluded to particles and plenums of energy that surround, encapsulate, and construct the subtle energetic foundation (*aura*) for material faculties, senses, species, bodies, action, intentions, and building ecosystems, societies, infrastructures, and civilizations—*human and animal.* Both Quantum and Ancient science (and even common sense) allude to a cause, intention, or need for anything to exist—the Hegelian Triad of *"Thesis, Antithesis, Synthesis"* is the dialectical formula. The game of Hide-and-Seek in the Puranas, like the elusive electron, is the newborn dialectical ability to measure the unmeasurable. That is Maya, and Quantum, the attempt to measure the infinite through reductionism.

Entering the material expanse, we reduce the manifest prisms of light and the surrounding vibratory commotion into visual-auditory interpretations of that very elusive measurement of the unmeasurable. The new expanse insists on reducing everything that is infinite into finite units of measure to accommodate the limited perceptual illusion. Whether we, the *Jiva-Atmans*, are photons of light or Electrons, Bosons, or Quarks is insubstantial and just as unmeasurably unimportant—to the infinite degree. Such comparison would be mere semantics. What matters, however, is the similarity in *"Mechanics."* Both ancient and modern speak of particles and waves from energetic and vibratory sources, playing the game of Hide-and-go-Seek, creating phenomenological universes.

Over time, 'thinkers,' philosophers, and scientists repeatedly studied the Ancient Texts and deduced similar subatomic findings to modern Quantum observations. In the end, both address the soup-like compound solution of substances and plenums inter-mingling and permeating the manifest universes. *"Substances," "plenums," "fields," "waves,"* and in the Puranas an *"Ocean of Causality and Consciousness,"* seem to be the consensus among thinkers, philosophers, and scientists.

> *Here the word 'ether' is used to translate the Sanskrit word 'Akasha.' Akasha is space, but it is considered to be a substance or plenum, rather than a void."*
>
> ~ *Physicist Richard L. Thompson, Ph.D. Mysteries of the Sacred Universe*

Professor Thompson continues:

> *Each element is regarded as the previous element plus an additional property contributed by a subtle sense element (tanmatra). Since the elements are derived from modes of sense perception, they appear to be, in a sense, insubstantial and to be ultimately based on consciousness. Thus from this Puranic perspective, the idea of decoupling an object from other matter and moving it through the ether is much more plausible than it is from the standpoint of modern thinking."*
>
> ~ *Physicist Richard L. Thompson, Ph.D. Mysteries of the Sacred Universe*

(It may be necessary to point out that the "accidental" death of Dr. Richard L. Thompson was never explained or investigated, with rumors of foul play)

...

The Substantial Nature of Consciousness-Driven-Substances (Realities) as Perceived Throughout Time

" *Ether is a **substance** filling every nook and cranny of the cosmos"*
~ *Albert Einstein*
(The Speed of Light: Constancy and Cosmos By David A. Grandy)

" *Everything is a "**substance.**"*
~ *Srila B.R. Sridhar Swami*

" *The mind (any mind) is a **substance**, and it is possible that it exists without the body (any body), and the body (any body) is a **substance**, and it is possible that it exists without the mind (any mind)."*
~ *Descartes' Conception of Substance*

" *Earth, water, fire, air, ether, mind, intelligence and false ego— altogether these eight [**substances**] comprise My separated material energies ...there is a superior energy of Mine, which are all [the] living entities [jiva-atman] who are struggling with material nature* ***and are sustaining the universe.***
~*Bhagavad Gita Purana 7:4-5*

" *God is a **substance** in the sense that he depends on no other entity for his existence, and created **substances** are **substances** in the sense that they depend only on God for their existence. Thus created **substances** are not fully independent, but their relative independence is enough for them to count as **substances** in an extended sense."*
~ *Descartes' Conception of Substance*

*"Aristotle held that the universe was divided into two parts, the terrestrial region and the celestial region. In the realm of Earth, all bodies were made out of combinations of four **substances**, earth, fire, air, and water, whereas in the region of the universe beyond the Moon the heavenly bodies such as the Sun, the stars, and the planets were made of a fifth **substance**, called quintessence."*

~ Numerous sources

*"Aether is a substance endowed with inertia, and capable, in accordance with the established laws of motion, of imparting its motion to other **substances**."*

~ William George Hooper, Aetther and Gravitation p.77

*"Somebody seeks out silver, somebody seeks out gold, somebody seeks out mica. This is knowledge of gross things—the earth. If you go to finer **substances**, then you study water, or liquid things, such as petrol and alcohol. Go still finer, and from water you will go to fire and electricity. If you study electricity, you have to study all sorts of books. And, from this finer fire, you will come to air. Now, we should inquire into the existence of the ego [ahankara], the finest material **substance**. What is ego? I am pure soul, but with my intelligence and mind I am in contact with matter, and I have identified myself with matter. This is false ego. I am pure soul, but I am identifying falsely. For example, I am identifying with the land, thinking that I am Indian, or that I am American. This is called ahaṅkāra. Ahaṅkāra means the point where the pure soul touches matter. That junction is called ahaṅkāra. Ahaṅkāra is still finer [subtler] than intelligence."*

~ A.C. Bhaktivedanta Swami Prabhupada

		GROSS	TRAIT	ABILITY	ACTIVE	
		substance	objects of the senses	receiving sense organs	working senses	
intelligence		earth	odor	nose		
		water	taste	tongue		
mind		fire	form	eyes		
		air	feel	skin		
ahankara [ego]		ether	sound	ears		
jiva atma						

All these **substances** are strewn everywhere unmanifest, (pradhana)
until touched by Time, due to the desire of the jiva-atman.
(Illustration by the author, James Ordonez)

*Cascading sambandha (relationships)
transform to action (karma) & substance (vibration)
creating acquired dharma(s)
for measurable existence (Maya)*

Vishnu - The maintainer of the Universe, also sitting in the heart
of all living beings as the Life-Force and companion
(Photo-Illustration by the author, James Ordonez, from personal photos)

Chapter 6

Yes, Karma Yoga Sounds Easy

" Only the UNKNOWING speak of
karma-yoga as being different from the analytical
study of the material world. ...[either] one of these
two paths achieves the results of both.
~*Bhagavad Gita Purana 5:4*

Now that we have a *'glimpse'* on the Puranic science of *sambandha* as a meditation to continue our journey, we can lighten our load—just a little. These are some heavy topics, especially when in the Western World, our points of reference differ. That is both true in our Western upbringing and apparently deep within our genetic memories.[89] In the Western World, both are intrinsically light-years away from the Eastern way of thought. A whole new topic, therefore, develops.

---

89   *Genetic Memories – http://bit.ly/Genetic-Memories*
      Phys.org

In the previous chapter, we introduced the analytical Puranic study of *"Modes of sense perception,"* with the help of Professor Richard L. Thompson. Once one understands this new vernacular, this science of the Puranas is entirely holistic and genuinely logical. If for no other reason than to give it a name and association, we chose for our discussion to also call this vernacular *sambandha.* After all, this is synonymous with that science of universal dialectical *Relationship* discussed; therefore, the label is appropriate for our purposes.

The Puranas continue. The *Three Modes of Material Nature* are described in the Bhagavad Gita and other Puranas, as follows. These are part of the ecosystem of the "consciousness-based" universes and creation.

> "Material nature consists of three modes—
> goodness, passion and ignorance.
> When the eternal living entity
> [initially] comes in contact
> with nature, s/he becomes conditioned
> by these modes."
>
> ~ *Bhagavad Gita Purana 14:5*

This initial *contact* begins back at that fork in the road we discussed where the *first foreign substance to the soul* is manifest by the *jiva-atman* (living being). There, also is, where the development of various consciousness-based energies begins to form an ecosystem. Based on *ahankara*, the material ego, through the Puranic notion of dialectical transformations—we can also begin to introduce this paradigm as a veritable *Ego-System*.

Here, the word *"modes"* is used as in *system, process, mechanism, or modus*. Since everything is based on consciousness, some may argue that the development or *emergence* of our ever expanding creation is paramount to an Operating System—a universal computer of sorts. Instead of binary combinations of commands, binaries, or Qubits, our system is built with *thoughts, desires, actions, and reactions*, looping *Ad Infinitum* in infinite directions.

> *Space and Time is a form of thought..."* [90]
> ~ *B.R. Sridhar Swami Maharaja*

According to the Puranas, the mechanics of this karmic law are strewn *throughout* all material existences. They simultaneously constitute the foundational field of energetic existence that compose the material universes. This *"Ego-System"* becomes the vast and infinite reciprocity of endless *Relationships* between all things, forces, and energies driven by intention. They are merely and magically just *thought*. According to an impressive variety of philosophers and thinkers from distant and recent history, *"space and time are synonymous with thought."*

---

[90]   *Not part of the "The Cult" Swami Tripurari goes deep - http://bit.ly/2OLCdax*
   Harmonist | Anadi for Beginners: We All Have to Start Somewhere...Or Do We?

As we have discussed, *Karma* in the original Sanskrit definition literally means "*action*," and nothing more. Understanding these Relationships and meditating on their cosmogonal splendor is called *Sankhya Yoga*.

Karma-Yoga, on the other hand, refers to pious activities while being aware of and meditating on the *sambandha* relationship with all things. Simply said, *Karma-Yoga* means to connect with God through "Goodness" with pious activities to other living beings, but with the aforementioned awareness of the connectedness of all things. Simple acts of kindness or piety without *sambandha* merely give one a good reaction without any yoga benefit of union with Universal Energies—according to the Puranas, of course.

> "*Karma Yoga is the taking of the things you do every day with other people, of service, and making those all into an offering. And so it's an attitude that one has.*"
>
> ~ *Ram Dass*

The various allusions through the Puranas indicate that *Karma Yoga* alone without the *Sambandha* understanding from *Sankhya Yoga* cannot be maintained. Yoga means to join or yoke with all things—*the Complete Whole*. In that same spirit of relationship and dependency, all the forms of yoga are *integral* and essential for the true Yogi altogether.

In *Karma-Yoga*, there is also an element of *Bhakti* (devotion). Many consider *Bhakti-Yoga* to be higher than *Karma-Yoga* because the sense of "devotion" is explicitly dedicated to God's deity directly and exclusively. However, *Bhakti* alone then becomes dogmatic, sending all the ubiquitous parts and parcels of the *whole* (everyone else) to the Proverbial "back of the bus." Somehow, in our experiences, these groups tend to upstage the Puranas' real message with *"religious Literalism."*

Without grasping *Sambandha*, or *Sankhya Yoga*, then *Bhakti Yoga* also falls flat. Once one grasps the teachings of the holistic Puranas, the clear message is that the ancient texts were not meant to separate or divide but to celebrate the unity of all things, from all angles. Arguably, that would be *Bhakti's* better definition, to embrace (or worship) the Complete Whole with devotion (not only on a physical altar).

This exclusivity preference is where the Puranas' Science becomes just another "Religion", bringing with it all the dogmas, sectarian, cultural, and judgmental *"Ego-System"* characteristics. Therefore, the entire process of *Bhakti* is compromised simply for being deemed "higher" or better than others by its practitioners.

Their sectarian mindset proposes that one can "only" connect with God through *Bhakti Yoga*, but not *Karma Yoga, Sankhya Yoga,* or any other spiritual way—whatsoever. This is a colossal snare among and between very similar groups of thought, cults, or yoga philosophies. All agree that everything that exists is *"simultaneously one and different"* with the Godhead. Of course, each insists that theirs, however, is superior. *(Hmmm! Now you see why we call them knuckleheads?)*

...

Some schools argue the Taoist point that "the more one tries to catch the *Tao*, the further away, the *Tao* vanishes." *[Funny that this sounds like the elusive electron]* This, of course, raises the age-old question in these teachings, "If God is everything, everywhere, and in everything, why can we not see God in our neighbor or even the objects around us and give them the same reverence and care?"

> *Without a solid grasp of Sambandha all one's bhajan (attempts†) remain on the neophyte level."*
> ~Guru Bhaktivinode

† Bhajan means "offering reverence" and loosely, "prayers or meditations".

In any case, the notion of any Yoga without all the others becomes another form of the *illusion—Maya*. The holistic approach to studying the Puranas is all-inclusive, focusing on the relationship of all things altogether. Therefore, this then becomes more of a science of The Complete Whole rather than a religion. The indication from the depth of these more profound messages of *Sambandha* is that faith, dogma, and sectarianism are the very antithesis of any universal or godly relationship. Religion becomes a competitive demon of sorts as a subset *"Ego-System"* formed by various tribal relationships—driven by the *Three Modes of Material Nature.*

Enter Cause and Effect!

Without all the noise and clutter of *value-competition* between doctrines and cults, what is left is the observations of thinkers and sages who, over time, stitched together the Puranas' hidden messages. Across eons, ancient Yogis considered these texts the Science of Creation—not a tribalistic sectarian cult.

Without all that distraction, we come back to the Puranic basics that "Space and time is a form of thought," and that everything is based on *consciousness.*

Elemental consciousness, or presumably *"decoherence,"* on the universal scale, is an organic ecosystem of desires and thoughts manifest into intention and vast interactivity. Overall, the universe is a living, breathing organism[91] composed of many living *Ego-Systems* inward into the microcosm's eternal abyss, and ever-outward to the unending macrocosm—ad infinitum. This is the premise of the *all-pervading consciousness* that the Puranas allude to being the real meaning of *"Infinity."*

> *The universe is a single living creature that encompasses all living creatures within it."*
> ~*Plato*

'Cause and Effect' is not so much a 'law' but a symptom of action—*karma*. The Yoga of Karma is, therefore, based on understanding the cause and effect of one's actions, intentions, and thoughts. With this understanding, the general definition of *Karma Yoga* is explained in Wikipedia.

"Karma yoga is a path to reach moksha (spiritual liberation) through work. It is rightful action without being attached to fruits or being manipulated by what the results [value] might be, a dedication to one's duty, and trying one's best while being neutral to rewards or outcomes such as success or failure."[92]

> *Selfless action performed for the benefit of others."*
> ~*James G. Lochtefeld, Ph.D.*

---

91   *The Living Universe – http://bit.ly/The-Living-Universe*
      Huffington Post
92   *Karma Yoga – http://bit.ly/The-Yoga-of-Karma*
      Wikipedia "Karma Yoga"

Of course, the Puranas goes deeper than that. *Bhakti Yoga* aficionados will dismiss *Karma Yoga* as essentially very inferior for being philanthropic and motivated by *"spiritual liberation."* The misconception is that doing good for others rather than ritual services to the temple's deity is a "selfish act," in comparison. This critique assumes that all Karma Yogis are only in it for the value goal of liberation—not because of empathy or compassion for others, or even their love for God's creation and all living beings.

However, the fact is that the same Puranas that speak about *Bhakti Yoga* and *Karma Yoga* also reminds us that every subatomic particle and every living being are simultaneously one-and-different with their source creator. The cause of all causes is ubiquitously manifest everywhere. This all-pervading presence is the ultimate deity of all creations, material and spiritual—*Vishnu* in the form of the Super-soul or *Param-Atman*, also known as the Lord in the heart.

If a sense of devotion is intended towards this form of Vishnu in the heart, the Karma Yogis may argue that *Karma Yoga* is, in fact, *Bhakti Yoga*. They may also argue that this devotional form of Karma Yoga is sufficient and equal—without the altar, the temple, the group *(sangha)*, social expectations, the dogmatic taboos, the cult, and the cultural, political, and *economic entanglements*.

> "Karma Yoga really is serving others as a way of serving God. You serve others as a way of putting flowers at the feet of God, honoring God, and so doing your 'seva,' or service."
>
> — *Ram Dass*

> "*In the simplest sense, you could say that Karma Yoga is using your karma as a way of coming into 'yog', or union with God, by using the 'stuff' of your life. Using it as the way in which you do work in the world, and acknowledging whether or not that work in the world is a vehicle for spiritual awakening. So doing your 'seva', your service, can work in a devotional sense, where you are consciously considering your action as an offering, saying, 'This is my Karma Yoga, I am doing this now as service to you, as an offering to God, and it's work on myself.'*" — *Ram Dass 2019*

With a "*Solid grasp of sambandha*," the difference between *Karma Yoga* and *Bhakti Yog*a bridge over as one—*sambandha* is the cohesion. Without a "*Solid grasp of sambandha*," both remain incomplete.

The trick with *Karma Yoga* is to learn to see the ubiquitous deity of creation—*Vishnu*—in all situations and within all living beings.

This is a much more loving and easy formula to follow for many folks, without exclusivity and sectarian division.

...

Here's where future societies may need to go!

Without religion at all, however, *Karma Yoga* stands alone, actually superior—without the racist (bigot) boundaries of sectarianism and supremacy that divide us all. If we were to educate ourselves on New Age Quantum-like Karmic reasoning, that everything is a *Universal Consciousness* everywhere without borders or labels, *Karma Yoga* can be practiced everywhere without religions, cults, or boundaries. Here is where this confluence of modern Quantum and ancient Puranic reasoning may press us to consider an educational requirement for K-through-12+ in society.

Perhaps it's time to make general populations aware of the Quantum and Karmic Dialectical conundrum with the Mental Mechanical Universe?

 Schools, to this day, begin by teaching children classical physics, glorifying failed past History (with lies about Columbus and Thanksgiving), racial separatism in religious schools, and proven failed *dialectical* economics.

It's like the idea of money (currency). It's both an illusion and a *dialectical substance*. Conceptually, money is based on the value of a tradeable item like gold (which is nothing but a "substance" measured by its scarcity)—conceptually. But, it's a *[bleeping]* lie because there never exists substantially enough quantities of the scarce substance to justify any measure of value. Whenever such dialectical value comes close to any actual measurement, it changes in value, and the value of the currency fluctuates as a result—*the Dialectical Triad*. Yet, our survival is dependent on this lie. Countries have milled coins and printed currency of all sorts since barter began in the Stone Age. Still, the dialectical substance idea of measure and value from the over-inflated scarcity of another substance (with no purpose) makes us all idiots and slaves to a dialectical global economic cult. Now we have BitCoin, a dialectical argument of value based on the buying and selling of another *dialectical idea*—proving the invalidity of gold—yet creating another dialectical illusion of tradeable value—and a new currency cult has emerged from the mass mind.

• • •

Back to our "poster child"

The modern-day problem with *Karma Yoga* is "The Cults" attempting to minimize *it (as a value) over Bhakti Yoga* for self-aggrandizement, donations and profits. This behavior is typical of tribal cult mentality. While concepts may be logical, to most, and corroborated by notable thinkers, a "Cult" group mentality will need to invent some ownership for vanity. For example, while *Karma Yoga* is predominantly emphasized in the Bhagavad Gita with an entire chapter devoted to it—albeit with a sense of devotion—*Bhakti Yoga*'s superiority needs to be constantly stressed by "The Cult" for self-admiration. However, doing so is the antithesis of *Bhakti Yoga*, precluding other forms of worship over one's own. Cults, whether political or religious, follow this same pattern of *value and exclusivity*.

That *exclusivity*, again, is from the acquired *temporary dharma nature* of the tribal and herding genetic instinct, as the dialectical encoding of humankind's subliminal unconscious bias.[93]

In brief, *Karma Yoga* and *Bhakti Yoga* are predominantly synonymous in the *sambandha* relationship, simultaneously the same and different—with only a slight difference of the vehicle perspective invoked. In *Karma Yoga*, the vehicle is kindness and empathy for others and the world, as God's creations. One could argue that *Bhakti*, whose vehicle is the deity on the altar, concentrates

---

93   *Tribal Herding Genetic Instinct – http://bit.ly/Tribal-Herding-Genetic-Instinct*
        <u>Harvard Business Review</u>

too much on the practitioner's well spiritual being, precluding empathy and kindness to others.

As we have alluded to before, the principal edict in the *Puranic* theory of relativity—*sambandha*—is the essential foundational platform of creation known as *achintya-bheda-abheda tattvam (literally meaning identical and different platform)*. This "simultaneously one-and-different foundation" of God and *Their* creations is the magic behind all relationships and interactions. The foundational Sanskrit word *tattvam* is best defined as 'a principle, or state of creation'—a *substance, plenum, or "field."*

☞ *Please note that we used the plural possessive pronoun "their,"[94] for the Gods, as we do in modern times to indicate members of the LGBTQ community who are either genderless or both genders. We do so here to indicate that God, in fact (according to the delivery of the Puranas), represents both genders while being genderless and simultaneously both genders in their own majestic universal magic.*

While modern science offers many theories, the underlying common thread between the various scientific theories and the Puranas views is the existence of an energetic, dynamic and interactive "field" or fields pervading all creations—a form of universal consciousness—as an active principle that animates creation.

----

94   *Gender Neutral Pronouns - http://bit.ly/Singular-They*
      Dictionary.com - Gender Neutral Pronouns

But first, we must explore the multifarious true meanings of *AUM (Om)*, and how integrally this presumed divine vibration holds and facilitates everything—altogether. Another essential *sambandha* relationship without which the puzzle is never complete.

...

☞ *KEEP TIGHTENING THAT SEATBELT*
...This just keeps getting stranger...
(in a good way)
STRAP IN TIGHTER!

...

Remember...
there's gonna be a quiz.

Durga - The shakti of war, strength, and protection
Photo-Illustration composite by the author, [95]
James Ordonez from personal photos of the Metropolitan Museum of Art

----

[95]   *Durga – Metropolitan Museum of Art - http://bit.ly/Durga-Metropolitan-Museum*

Chapter 7

The Implications With OM

AUM bhur bhuvaḥ svaḥ 'TAT'

O m, or *AUM,* is of paramount importance in Hinduism. This symbol is the sacred syllable representing *Brahmán,* the *impersonal* all-expansive, all-pervading Absolute. In the material expanse, the *Brahmán* aspect of the Absolute is *ubiquitous, omnipotent, omnipresent,* and the source of all manifest existence. This *Brahmán* feature differs from otherwise many demigods and *shaktis* (energies or substances) that are often worshiped, thought of, or meditated upon as individual deities—they are one and many—*"simultaneously."*

It was Einstein, Descartes, and Aristotle, among many, who began to agree on an operative definition of nature's forces and energies using the term *"substance."* The Puranas also explain that everything is a *"substance,"* brought into *coherence*[96] from *Om's* original vibration (composed of three separate syllables in Sanskrit, *A, U, and Ma*). Before space and time as we know, pure consciousness existed completely "unadulterated" within the primordial realm of origin and causality—beyond the material creation. This realm explained as a state of pure consciousness, is in the Sanskrit language is described as *"Sat-Chit-Ananda—ever blissful, all cognizant eternal infinity of timeless existence."* The Puranas' description indicates a world or domain beyond matter, beyond temporality, and most importantly, beyond ignorance and the duality of suffering—*the all-knowing non-dual*[97]. That is Brahmán.

> **"** [Brahmán] the creative principle which lies realized in the whole world"
> ~*Wikipedia*[98]

The Puranas explain. The vibration produced by chanting *AUM* in the physical universe corresponds to the original vibration that first entered the material realm at the time of creation. The

---

96   CO·HER·ENCE – *the quality of being logical and consistent and forming a unified whole*
97   *Advaita Vedanta - http://bit.ly/Advaita-Vedanta*
       "The term Advaita refers to the idea that Brahman alone is ultimately real, the phenomenal transient world is an illusory appearance (Maya) of Brahman, and the true self, atman, is not different from Brahman."
98   *Brahman – the Ultimate Reality in the universe. - http://bit.ly/Brahman-Ultimate-Reality*
       Wikipedia – Brahman

sound *AUM*, also called *Pranava*, meaning that *substance* that sustains life and runs through the *Prana* or breath. *AUM* also represents the four states of the Supreme Being, as follows. The three individual sounds in *A.U.M.* represent the waking state, dream state, and deep sleep states, and then the fourth, the silence surrounding *AUM*, represents the state of rest.

Whenever *AUM* is recited in succession, there is an inevitable period of silence between two successive *AUMs*. This silence represents the fourth state known as *"turiya,"* which is the state of perfect bliss when the individual self recognizes his identity with the Supreme Consciousness Persona. Only in that state of *turiya* can one understand or perceive the magic behind *karma*.

> "When the highest type of men hear Tao, they diligently practice it. When the average type of men hear Tao, they half believe in it. When the lowest type of men hear Tao, they laugh heartily at it. Without the laugh, there is no Tao."
> ~ *Lao-tzu*

In Zen Buddhism, the teachers give paradoxes or impossibilities (*koans*) to the students to meditate upon, like the famous "sound of one hand clapping." That perpetuity of the paradox or contradiction fills the intellect. The meditation reveals the paradoxical, insoluble puzzle of existence to the heart—while not allowing other thoughts to seep in (much like mantra meditations), the mind becomes a captured audience. A great book to read is Roshi Philip Kapleau's *The Three Pillars of Zen*.

Recommended Read & YouTube Channel
The Three Pillars of Zen '

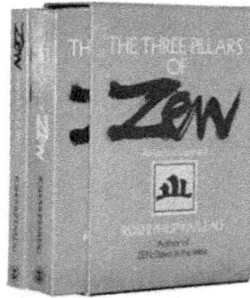

The Three Pillars of Zen By Roshi Philip Kapleau [99]
(Photo and retouching by James Ordonez)

The Puranas continue. Before the beginning, before Time, the very source and essence of *Brahmán*—known as *Krishna*[100]—deep within His Persona, was one and non-dual. That *Consciousness* Persona, the 'Absolute Reality' thought, "I am only one, may I become many." This caused a commotion, a vibration that

---

99   Roshi Philip Kapleau – Recommended YouTube Channels – http://bit.ly/Three-Pillars-of-Zen –
      YouTube Recommended Channels
100   Definition: KRISHNA – http://bit.ly/Definition-KRISHNA
      Krishna, literally "attractive", akin to dated Russian красная -- "beautiful", also similar to chrysos meaning "gold" in Greek is the eighth incarnation of the supreme god Vishnu in Hinduism. The word Krishna means One With Dark Complexion and One Who Attracts All.

eventually entered the material realm and became sound, and this sound is *pranava: AUM.*

Creation itself was set in motion by the vibration of *AUM.* The closest approach to *Brahmán* from within the material expanse is this very first sound vibration. Thus, this sacred symbol has become emblematic of *Brahmán,* just as images are representations of material objects to our eyes.

The first of the three states of consciousness is the *waking state*; it is represented by the letter *"A,"* pronounced as in the word 'apple.' The *dream state of consciousness* lies between the waking and the *deep sleep states*; it is represented by the letter *"U,"* which rests between the *"A"* and *"M."* This *"U"* is pronounced as in the word 'soup.' The last state of consciousness is the *deep-sleep state* and is represented by the letter *"M,"* pronounced as in 'zoom.' This third syllable closes the pronunciation of *AUM,* much as REM sleep enters the final stage of the mind at rest. This silent rest, which follows each iteration of *AUM,* represents the *Turiya state of perfect bliss.* The living being then recognizes the peace that "not a speck of dust in all of the universe is out of place"—in harmony and complete.

Therefore, *AUM* is the Original, Perfect, and Complete principle of All that is. These four vibrational forces *(including the silence between)*, all of which existed before time and space, are the vibrational foundation of all material creation. The silence after *AUM* is the Absolute Consciousness and essence that permeates the other three, as the *Source*, *Origin*, and *Personality* of all existence. This peaceful silence is the *sambandha* relationship of the living being with the ultimate state of pure consciousness— the Persona of the Complete Whole—by whatever name of "God" is used.

Joseph Campbell quoted the Hindu scriptures that he translated from the *Sanskrit* language concluding that "truth is one, the sages speak of it by many names." However, the conundrum brought in by humankind is the tribal tragedy of sectarian, racial, religions and cults. They narcissistically preclude and even demonize all others who do not adhere to their chosen symbols for God or Truth.

With this limited and judgmental understanding—from the world's scriptures and texts—they distribute whatever skewed interpretations work best for them, often carelessly and irresponsibly. "Flowery words" from the many world scriptures become rhetoric and empty promises for building entire empires based solely on monetary donations from innocent congregations. Small philanthropic *façades* and spiritual assurances inadvertently create wealth, power, and avarice, skewing the well-intentioned spirituality into *business enterprises—just like political cults!* That is inevitable by the acquired nature of humankind and the dialectical consciousness-driven *dharma*. It's quite a circus.

> "The unwise, of small knowledge, are very much attached to the flowery words of the Vedas [and] say there is nothing more than that"
> ~*Bhagavad Gita Purana 2:42*

Therefore, when discussing these spiritual topics of *AUM, sambandha, shaktis*, etc., one has to be aware of the presenters' skews and affiliations, cults or separatist agendas. In our journey through the vast Puranic perspectives, we learned that one could argue just about any point whatsoever using the "Flowery words" of the Vedas, Puranas, or any ancient writing. After all, these Texts have been gradually compiled, passed on, and rewritten literally over eons, through many different cultures, civilizations, and situations. There exists an anecdote or parable for just about every possible situation or problem. This known factor, along with the "Religious Literalism" described in Hermeneutics' studies, allows humankind to extrapolate falsehoods. Our minds, exceptionally as *mass-mind* in groups, can propose any possible Twilight Zone skew from the writings in ancient Texts *or* even the current events of the last few centuries. These phenomena are not limited to the Puranas and the Vedas. It also extends to the Quran, the Bible, the Torah, etc. Modern-day religious cults, political cults, and conspiracy theories fall prey to this despite scientific advancement and progressive modern thought. It's almost like humankind is genuinely not very bright despite all the Ph.D. and educational posturing flying around. A group can come to believe and proselytize that the sky is red, the Earth is flat, or that seventy-two virgins are waiting for us in heaven if we destroy non-believers. Or, like the QAnon[101] Republican QOP party of the United States believes, that "Jewish space lasers from outer space are causing California wildfires.

•••

☞ *Our "Poster Child" examples,*
"The Cult Branch" and "Mr. Swami" in question,
help us identify the dynamics and results of cults and separatist,
discriminatory agendas!

These groups and cults actually do the most considerable disservice to the truth, equality, fair play, and veil over with *Maya's* illusion. An example is the religious cult in question, "The Cult Branch." It began when Swami Bhaktivedanta graciously and with selfless and humbling sincerity came to the West in 1965, disseminating what he literally called *"The Science of Krishna Consciousness."*

In his dozens of published Puranic translations, Swami Bhaktive-

---

101   *What Is QAnon? – http://bit.ly/What-Is-QAnon*
       <u>The New York Times</u>

danta monumentally detailed this Puranic science of *brahmán*, AUM, *sambandha*, and the cascading *24 substances* of creation. He carefully explained their entrance into the material realm. Sadly, his translations and purports were edited over the years, skewing the message here and there—*an Oxford Comma here, a capitalization there, a synonym over there, or simply better grammar over intended meaning.* Suddenly, the Earth was physically flat, with little regard to the Texts indicating a "plenum," "field," or level of *shakti* energy. Can this be Selective Reasoning or Motivated Reasoning, or both?

"A camel is a horse by committee!"
~an old saying

Initially, in his many purported Puranic translations, Swami Bhaktivedanta consistently and ecumenically declared over and again the essential need of seeing the ancient texts '*as a science*' and not just a religion. In fact, one of his many famous yet vastly ignored edicts was, *"Religion without science is sentimentality and science without religion is speculation."* He was referring to the Science of the Puranas, *sambandha*, not classical physics.

In the end, every cult has its own *market-positioning* messaging spin, holding their respective audience captive—like any competitive advertising or marketing campaign. The group becomes just another business enterprise like the many Christian denominations, neither responsible for taxes or accountability—encouraging unethical and even criminal behavior. They hide behind Godliness for job security and career.

Years back, a "President" of the Los Angeles "Cult" temple actually said to us, "Do you realize that if I just go to the morning program *(services)* daily, I will have job security for life." This was actually said without shame, as if proud of their career accomplishment. Reportedly, temple presidents in "The Cult" earn up to six figures yearly, *(illegally?)* from donations, "off the books," while incumbent hard-working cult members receive only three meals a day and a place to sleep. Such "career" mentality we found just about everywhere in their groups' leadership with few exceptions. At the very top, a cult's leadership, motive, and intention are typically and by default driven by survival instinct, comfort, and achievement. They *are just people*, taking advantage of their position, as in political cults. The really sad part is that they do not see it—as far as they're concerned, they've convinced themselves that they are *"poor [bleeping] mendicants doing the best 'holy' they can."*

Such business-like mentality upstages and destroys the *religious cult's* original intention—*"spirituality."* The paramount importance of *Pranava AUM, sambandha,* the *24 substances, shaktis,* etc., and this exact science of the Puranas brought by Bhaktivedanta Swami, are, therefore, for the most part, ignored by *"The Cult."*

Some scholars linger here and there obsessed with the ancient Puranic science, but those are "far and few between." Then there are other so-called "scholars" preoccupied with teachers' and saints' genealogies, but mainly around the premise of religion rather than the Puranic science. Both are careful to stay within "The Cult's" guidelines of what can and cannot be said. Else, the authors and their books are shunned or forbidden by "The Cult" and their book stores. (*We are confident our book will not be displayed in their temple bookstores.*)

☞ *Once we tie this all together, and with all the 'celibate sex' you may agree, we could make a very funny Sci-Fi-Action-Dramedy Soap mini-series!*

On the whole, these *"essential"* scientific Puranic nuances and teachings are merely mentioned in passing in "the Cult's" readings and lectures. In fact, these are instead taught to be inferior and of less consequence or need than just and only worshiping the deity on the altar and furnishing the temple with monetary donations and free volunteer labor. Sound familiar? The excuse is *Bhakti Yoga*, which also falls prey to, and is misinterpreted, the same way as mundane politics, money, and human avarice. We will dive deeper into authentic *Bhakti Yoga* later on.

What Swami Bhaktivedanta brought to the West was, therefore, quickly and efficiently cannibalized by his own followers—almost immediately after his passing from this world twelve years later. All his "disciples" began fighting with each other, especially the leaders to this day. They split into the above-mentioned splinter groups.

The takeaway is that the illusory material energy of *Maya* is infinitely complex and difficult to comprehend. Self Realization does not happen simply by being intimately associated with a "spiritual" group, especially if the group is engaged in questionable breaches of ethics, empathy, and compassion in the name of their mission to save the world.

> "The intricacies of action are very complex and difficult to understand. Therefore one should know properly what is action (karma), what is inaction (akarma), and what is forbidden action (vikarma).
>
> ~ *Bhagavad Gita 4:17*

...

Nevertheless, the Puranas indicate clearly that the seed of all material vibrations is the syllable *AUM*. We also find this invocation at the beginning of many mantras across various schools of thought. This *seed mantra* is also the cornerstone and foundation of the *sambandha* relationship of all things. It is also indicative of the "Cause of All Causes," known in Sanskrit as *Karana Karanam*.

Vishnu – The direct (one and the same) expansion avatar from Krishna, which forms the trinity presence of God in the Material expanse.
(Photo-Illustration composite the author, James Ordonez from personal photos)

Deeper still, we go into the various associations and meditations included in the powerful syllable mantra *AUM. Karana Karanam* refers to the *Ocean of Cause and Causality*—our aforementioned *"Ocean of Consciousness."* Happening, both within the material expanse and outside, *synonymously*, in the infinity of consciousness that pervades everywhere. That in itself is a complete meditation that many yogis choose as their lifetime attention and meditative focus. *AUM* is complete on their own.

The Puranas do explain—even though some "cult" pundits may insist otherwise—the above meditation and form of *bhajan*, or worship, is actually complete in and of itself. This *AUM* mantra meditation is more akin to the Hindu worship of the form of *Vishnu*—a trinity that is taught in the Puranas as a direct expansion of God into the material expanse as the facilitator, maintainer, and companion to all living beings. Within the utterance or remembrance of this three-syllable sound reside the three aspects of this Vishnu trinity. The many sects in Hinduism each select a central figure or personality from the many expansive energies (*shaktis*) expansions of the ultimate *Cause of All Causes—God Them-Self.*

We use this pronoun (*them*) delicately to mean both gender neutered, as well as the *Yin -Yang* polarity of male and female symbology, as understood for procreation and reciprocal engagement. The latter, a reciprocal relationship, is the very source of the *sambandha* we have discussed in previous chapters—where it all begins.

Some Hindu sects select the personality of *Lord Shiva* as their central meditation or worship, while others choose *Vishnu*, or *Lakshmi*, or *Durga, Parvati, Yogamaya, Ganesh*, etc. Each indicates and represents another sublime energy expansion or *shakti* from the Complete Whole of All Consciousnesses. There are variations in forms of meditation, practice, or worship. For some yogis, the act of silent meditation alone is their chosen complete path. Others prefer mantra (*bhajan*) meditation. For others, worship alone, as in reverential service and rituals, is the personal choice. Still, some indulge in the various combinations to not leave any stone unturned and find solace in that.

Among the Shaktis (avatars) expansions from the ocean of consciousness pervading all creations in the material expanse are Lakshmi the Goddess of Fortune, Sarasvati the Goddess of learning
(*Illustrations by James Ordonez, from personal photos*)

The magic explained in the Puranas is that these forms of *sambandha relationships* are all the same, and differ only in the subjective intention and spiritual wherewithal (*adhikar*) of the individual yogi performing the meditation, *bhajan*, or worship. The Puranas explain how these magical *shakti* energies expand from a central point, like the petals of a flower unfolding realities and energies, all in concert with each other in a whorl of sorts uniquely separate and yet identical simultaneously with the central corolla. This, in itself, is a complete meditation.

AUM begins there, at the central corolla from whence the expansions begin. It is stated throughout that there, beyond the material expanse, pervades the original trinity and Cause of All Causes. These are represented in various Hindu sects by varying names and even deities representing qualities or anecdotes.

For Example, in one Purana, the three Personas forming the complete trinity are described. First, the consort *(Hladini shakti)* Radha, without which the central Godhead figure Krishna is incapable of existing. Her love manifest is the central Godhead, *Krishna*. They are completed with his brother *Balarama*, who then handles all other matters. They multiply and branch out into infinite subjective expansions of their own shakti(s) to perform and facilitate varieties of existences, dimensions, and *sambandha* relationships.

In another Puranic narrative, *The Ramayana*, there are three central Godheads, *Sita*, *Ramachandra*, and his brother *Lakshmana*. In yet another, we find *Lakshmi*, *Vishnu*, and *Ananta-Sesha*. These triads go on and on and cascade into energetic *shakti* levels that some consider *demigods*. These include *Shiva*, *Parvati*, or *Ganesh*, etc. However, those who worship *Shiva* and *Parvati* will defend that their deity is the center and *"cause of all causes."* The lines begin to blur. In the end, there is always a trinity and a series of trinities that all expand from each other. This again becomes one of those paradoxical, insoluble conundrums that challenge reason and thought. Still, the conception of *AUM, dharma, sambandha,* trinities, and *the Magic of Karma*, altogether remain the cohesive and constant foundation of the Puranic School of Thought.

	S U B S T A N C E S			
Gross				
	GROSS substance	**TRAIT** objects of the senses	**ABILITY** receiving sense organs	**ACTIVE** working senses
intelligence	earth	odor	nose	
	water	taste	tongue	
mind	fire	form	eyes	
	air	feel	skin	
ahankgra [ego]	ether	sound	ears	
Subtle jiva atma				

All these **substances** are strewn everywhere unmanifest, (pradhana)
until touched by Time, due to the desire of the jiva-atman.
(Illustration by James Ordonez)

"achintya bheda-abheda"
simultaneous oneness and difference
— the mantra less visited

Buddha Siddhartha
(*Photo-Illustration composite the author, James Ordonez from personal photos*)

Chapter 8

The Power of Mantras

All things *in Life are* **Vibrations**"

~ Can you guess what great minds alluded to this?[102]

B uckle up!! This topic is a rollercoaster ride of paradoxes and fleeting realities like the elusive electron, which habitually appears and disappears in its never-ending game of *hide-and-seek*.

Over and again, this ancient Puranic science explains the importance of sound-vibration holding all material existence cohesive and interactively engaging—beginning with the vibration of *AUM* as the underlying foundation of all creations. Other ancient writings allude to a similar message, like the Gospel of John 1:1, which states, *In the beginning was the Word, and the Word was with God, and the Word was God.*" We can only speculate John's

---

102   *All things are Vibrations – https://bit.ly/Life-Vibration*
       https://bit.ly/Life-Vibration

realization to be similar to the Trinities of *AUM*, discussed in chapter 7. Nevertheless, beyond the various scriptural and philosophical concepts of an original vibration, there is also significant collaboration from both Classical Physics and Quantum Science. In essence, everything material is nothing more than vibrations, which are sourced from combinations of even subtler vibrations. A preponderance of qualified opinions over eons indicates we are all instruments in a tumultuous symphonic concert and rave, through time and space—all interconnected in the weave of sound. As Pythagoras cited:

> *Each celestial body, in fact, each and every atom, produces a particular sound on account of its movement, its rhythm or vibration. All these sounds and vibrations form a universal harmony in which each element, while having its own function and character, contributes to the whole."*
>
> *~ Pythagoras (569-475 BC)*

The Sanskrit word mantra initially translates literally to "mind vehicle." The root word *"Manas"* means mind, and *"tra"* means vehicle. Some say "mind-Instrument" The word also is used in spiritual groups as "mind freeing" or "to free the mind." In any case, indications from the Puranas allude that the mass-mind is ubiquitous and progressively creative. That at the same time, the individual-mind can travel the weave of the Universe and *influence* creations in its surrounding ecosystem.

Having gotten to this point in the book, you may be ready to engage the power of mantras. We touched on *adhikar* a few times in previous chapters—'*developing spiritual desire.*' This is yet another instance where the reader may begin to measure their own spiritual interest. Since indeed, you are still reading, you have already begun to qualify. There is no right or wrong; *adhikar* is a developed desire for spirituality by choice. Mantra Meditation is introduced in the Puranas to awaken and further develop such *adhikar* spiritual desire.

Entire volumes are dedicated to the "Power of Prayer," and Joseph Campbell also focused on the "Power of Myth." Both inevitably involve the mind in dialectical communication and the development of karmic intention in our lives and situations. The Puranas continue to insist, *everything is based on consciousness.* As discussed by so many prestigious philosophers, scientists, and thinkers, we need to accept that the universal mind and our individual intentions travel and subsist in the vibrations of the many shaktis *(energies)* surrounding us. The entirety becomes a veritable *"Ego-System"* of consciousness *(ahankara).* These ancient Texts also suggest an ability to harness the power of vibrations around us to influence, transform, and even astral-travel[103] through the power of mantras.

----

103   WikiPedia – Astral travel is ancient and occurs in multiple cultures – http://bit.ly/Astral-Traveling.
       http://bit.ly/Astral-Traveling

However, these types of yogic abilities are no longer as achievable as they were in past ages—and were rare even then. Before mechanization and the industrial revolutions, humankind was embraced by *organic*, natural vibrations all around. We were much more open, in previous centuries, to feeling the Earth beneath our feet and hearing the natural universal forces around us. We now travel seventy miles an hour making split-second decisions, aided by machines and computerized decision-makers, and bombarded continuously by artificially amplified radiations from satellites and communication devices and towers. We eat processed, unnatural foods; our minds are mechanized by constantly surging digital media and manipulative propaganda and advertising we carry in mobile devices with us *addicted*—often filled with lies and divisiveness. Patriarch con-man politicians and State-Media (FOX News) cleverly manipulate the masses into dull headed submission and division. Unregulated torturous slaughterhouse animal agriculture farms are taking over entire tropical forests. *(And no one gives a [bleep!])* Twenty-first-century humankind has experienced a dulling of the natural ability to hear, feel, and empathize.

It almost seems like the empathy gene never got the chance to go viral through humanity before we altogether spearheaded the destruction of the planet through avarice, lack of empathy for both humans and animals, and male-led religious wars. Interestingly, genetic science[104] may even indicate that Earth would have progressed better if the dominant patriarchy would have shifted to a matriarchy or at least a balance. In any case, humanity has, for the most part, lost the ability to tap into the mystic. Therefore, don't expect *perceivable* magical results from chanting mantras.

All is not lost, however. This deficit in human development does not mean that the individual should not at least try and take advantage of whatever universal benefit is still available in our Twenty-First Century afflicted condition. After all, it is individual participation that eventually creates a herding momentum. Perhaps some healing or reversal may be individually attained if applied with sincerity and mindfulness. Unfortunately, the math at a glance does not bode well for all of humanity transforming anytime soon, if we are not extinct in the next millennium. The allusions from the Puranas still give us clues to move forward.

According to these ancient texts, mantra meditation connects the individual to the expanse of vibrations around us, thereby quieting the mind, allowing it to drift free of thoughts, desires, and emotions. This first step is the beginning of a journey back to primordial origins through vibrations. Contemplating our *relationship (sambandha)* with sound becomes our repose, and

104 *Females 50% more of the empathy gene) – http://bit.ly/Genetics-Female-Empathy-for-Animals*
   http://bit.ly/Genetics-Female-Empathy-for-Animals

the via media—a middle way or compromise between dialectical extremes. You may ask, "What are these extremes?" They are *Fundamentalism* vs. *Atheism*, the two products of "Religious Literalism." This Middle Way (*sambandha* mantra meditation) is essential to ensure the seeker neither falls to fanaticism from the studies of Hermeneutics (discussed in chapter 2), or worst deny the existence of God entirely from watching the above fundamentalists make total *[bleeping]* asses out of themselves. The "mind vehicle" (mantra meditation), when properly administered with *sambandha*, creates the balance.

In previous chapters, we lightly introduced the Puranic magical conclusion of "simultaneously one and different" known in Sanskrit as *"achintya bheda abheda."* This premise applies to all creations, including this one—mantras. We are often asked, "Do mantras give you mystic skills?" Beyond "Freeing the mind" for meditative purposes, this question falls under this magical *simultaneity* duality mentioned above. The answer is both "yes, and no" simultaneously, depending on either microcosm or macrocosm perspective. The Puranas press us to understand this metaphysical duality and paradox. Therein lie many answers to many apparently unresolved mysteries.

One of the Upanishads included in the Puranas begins with the following invocation mantra. *"AUM purnam ada purnam idam."* The complete verse loosely translates as follows.

> *AUM (Om), the Complete Whole, is perfect and complete. Being completely perfect, all emanations therein—everything—are also perfectly complete wholes. Whatever is produced of the complete whole is also complete in itself. Being completely perfect, AUM maintains the complete balance."*
> ~ *Sri Isha Upanishad verse 1 - Invocation* [105]

Down to every speck of dust or subatomic particle, the Puranas insist on the complete presence of Vishnu *(AUM)* inside all things, existences, situations, atomic particles, and even sounds, concepts, and thoughts. This Complete Whole is both the source of every object and the object itself in simultaneity. Mantras are a combination of sound, concept, and thought, altogether a complete meditation, to be recited repeatedly for full immersion. In the *microcosm*, individual mantra meditation quiets and delivers the mind from states of wanton disarray and confusion by the mere process of relaxation. This peace gives the individual the opportunity and clarity for spiritual advancement.

---

105 *AUM (Om), the Complete Whole, is perfect and complete – Word for word translation*
     Word for word translation and purport

> *Pronounced correctly, OM has four syllables and is pronounced AUM. When chanted, OM vibrates at the frequency of 432 Hz — the same vibrational frequency found in all things throughout nature."* [106]
>
> ~Discover Magazine

In the *macrocosm*, however, as in the power of mass prayer, cumulative mantra recitation is alluded to with more *perceivable* powerful influences—when administered by large groups. For this reason, joining into mass-mind-established mantra meditation (or prayers) with millions of others around the planet holds more promise of efficacy and results. Sanskrit mantras, in particular, are described to be in tune with *shaktis* and karmic forces[107], which have been chanted on the planet for thousands of years. With the recent globalization of Eastern Thought, mantra vibrations are now more than ever, globally reaching into the ether and stratosphere for cumulative effect.

It's important to point out here that the "vibrations" of prayer and mantras do not necessarily need to be audible in the physical sense, as thoughts and intentions are also vibrations yet more powerful. Therefore, silent mantra meditations and prayers also carry the *intention* of the mantra—a *vibration* in any case.

---

106   AUM = 432 Hz – http://bit.ly/AUM-Frequency
       Discover Magazine
107   The Power of Mantras – http://bit.ly/Mantra-Power
       The Secret Power of Sanskrit Mantras

Vibration is the vehicle

As in the traditional definition of ecosystems, it is not difficult to imagine a universal interdependence of *consciousness*. Such is the interconnectivity of relationships between forces, vibrations, intentions, and reactions, which we call *sambandha*. The Puranic Texts alludes to universal consciousness strewn everywhere in various forms, starting with *ahankara*—altogether an *Ego-System*.

From these allusions of the Puranas and the philosophical and scientific proposals by some modern science, our takeaway is eye-opening. We, everything, and all things are interconnected in a web of intention and response—*cause and effect*. Loosely—the overall evaluation in this Puranic proposition is that the *substance "universal mind"* is the glue that holds the other material *substances* together. Then, *consciousness*, the creative principle behind intentions, actions, interactions, and reactions, creates and manifests realities, cultures, and civilizations. Of course, we are keeping in mind three main points. The first is that the *universal mind substance* is a transformation of *consciousness*. Secondly, the *Jiva-atman* spark of *consciousness* responsible is but a drop in the *ocean of consciousness—the cause of all causes*. Lastly, all these are, in a sense surfing the vibrational waves of all these energies and shaktis hither-and-dither frolicking in a cosmic dance. *Vibration is the vehicle.* Therefore, from this *"Ego-System" (ahankara)* perspective, the concept of creationism becomes a continuous, willful, spontaneous, dialectical emergence from all-pervading consciousness—*perpetually unfolding and blossoming*—magically.

Only for the sake of exploration and fully grasping the essence, this now brings up the more magical, less viable proposals on 'the power of mantras from the Ancient Texts. Quantum Physicist Richard L. Thompson, Ph.D. states:

> *Since the elements (of creation) are derived from modes of 'sense perception,' they appear to be, in a sense, insubstantial and to be ultimately based on consciousness. Thus, from this Puranic perspective, the idea of decoupling an object from other matter and moving it through the ether is much more plausible than it is from the standpoint of modern thinking."*
>
> ~Quantum Physicist Richard L. Thompson, Ph.D.

Still, the *Bhagavatam Purana* takes this one step further.

> *The yogi who completely absorbs his mind in Me, and who then makes use of the wind that follows the mind to absorb the material body in Me, obtains through the potency of meditation on Me the mystic perfection by which his body immediately follows his mind wherever it goes.*
>
> ~Bhagavatam Purana 11-15-21

This Puranic verse suggests that complete meditative absorption of the mind on God could provide the ability for the rest of the

Aura *(subtle body discussed in chapter 5)* to follow, thereby, with perfection in such practice, also teleport the material body with the astral projection of the mind.

☞ *OK, now this is some crazy Sci-Fi!*
Or is it?

Reading these, we were reminded of our Junior High School Quantum Science teacher, Mr. Pepitone, declaring on our first day of school that "this blackboard eraser could be something as improbable as a cow."

It is not that we will try this fiercely far-out idea and attempt to teleport through mantra meditation. Yet, we thought, "Why is it not at least possible, regardless of how improbable?" It was, after all, Quantum Science that gave Star Trek's Gene Roddenberry the crazy notion in the 1960s that teleportation could be harnessed by controlling energetic vibrations of atomic particles through the ether.

All this merely drives a point. Is it possible? Absolutely! In fact, teleportation is currently being used to transfer and store data and information in Quantum Computing.[108] Studies of Ground-to-

---

108   *National Science Foundation – Is teleportation possible? – http://bit.ly/Is-Teleportation-Possible-Yes*
        Is teleportation possible? Yes, in the quantum world

satellite teleportation of particles using Quantum Entanglement begun in the mid-1990s. Then, in 2017 scientists in China successfully "teleported a photon from the ground to a satellite orbiting more than 500 kilometers above."[109]

As with Quantum Computing and Quantum Science explanations of subatomic interactions, the Puranas gave us the science of vibrational interactions and relationships, based on *consciousness (Quantum Decoherence)*, of similar particles and waves discussed in modern science. But instead of receiving and grasping the intended message, we have read the ancient texts and got lost in the Hermeneutics of *Religious Literalism*. So, we are here to untangle the Puranic premises of Karma and **karmic entanglements**, comparing with accepted modern science—in this case, with the power of mantras.

The Puranas continue. Each of the **celestial shaktis** introduced in these ancient texts indicates an energetic potency behind each alluded personality existing in creation. Loosely, *Ganesh*, for example, is 'success 'and the 'remover of obstacles;' *Lakshmi* is the potency of 'abundance'; *Vishnu* the maintainer; *Shiva* is 'Time as the destroyer,' *Maya* the 'illusory energy,' *Indra* the 'sky,' *Sarasvati* of 'knowledge,' *Varuna* controller of the 'oceans,'

---

109   *Science Alert – http://bit.ly/HowQuantumComputersWork*
        How Do Quantum Computers Work?

and so on. The Puranas describe each persona as the predominating deity or power behind the various potencies and aspects of creation.

In Greek mythology, *Poseidon* is the controller of the oceans, *Athena* of wisdom, and *Zeus* of the sky. The persona epistemology ends up being important for understanding our relationship *(sambandha)* with cosmic forces from *consciousness*. Much as with all other world mythologies, the assignment of a persona was not only genius but also appropriate if, indeed, everything is based on consciousness—*life-force*.

When we first grasped these explanations of the *personalities* behind all universal forces, it was an awakening epiphany. This was a genius revelation and erudite strategy by *ancient sages* to communicate what would otherwise be impossible to get across to the masses from different cultures and over eons through the ages. The *sages* needed to describe the persona in each shakti, or aspect of creation, in a manner that all humankind could grasp, and also with the careful insistence that *life-force* was behind the curtain of all creation.

Mantras, therefore, address the personas behind the *shaktis* and the by-design intention of each. As alluded to by the Puranas, the operative word for how mantras work is *"invocation."* The presence of these discussed energies in the form of vibrations creates the via media—*middle way.* By indulging in hearing and repeating each mantra, we invoke the c*elestial shakti.*

Mantras alone, without the expectation of magic, a mere relaxing focus and intention are invoked for calmness, drive, and perseverance for the individual. A holistic medicinal resource is invoked and administered. This meditation, too, is complete on its own.

If we consider the possible magic alluded to by the ancient texts, we find that cosmogonal reciprocity in the universal dialogue—the *dialectical sambandha* discussed. The Puranas suggest that we are, in fact, able to communicate with the Universe and the *shaktis,* and even God, through the invocation of mantras, by tapping into the holistic vibration of the *Ego-System.*

Therefore, according to one's own subjective *adhikar* spiritual desire, only the individual can gauge how deep to indulge in mantra meditation. There is no right or wrong; only what satisfies one's spiritual thirst, which differs in each individual, to feel connected to the Universe. What happens after that is between the individual and their *sambandha* relationship with the *Complete Whole.*

• • •

Popular mantras and their purposes do abound, not only in Sanskrit, but Pali, Tibetian, Tamil, Hebrew, and even Cristian Prayers in various languages are, in a sense, mantras. A well-known mantra is the Compassionate Buddha's chant, *"Om Mani Padme Hum,"* which translates to "Hail to the jewel in the lotus." This mantra is said to calm fears, soothe concerns, and heal broken hearts. Another Buddhist mantra is the Mahayana mantra, *"Nam Myoho Renge Kyo,"* which asserts that anyone may have the ability to deal with and overcome problems in life and transform suffering into non-suffering. Buddhist mantras tend to be self-healing and compassionate.

Beyond "Freeing the mind," some interpretations of the Puranas allude to the ability to form spells and curses and the potential for attracting or dispelling energies and shaktis to and from one's own destiny. The Western past is riddled with stories and believes of *Voodoo* and *Santería* type spells and curses to manipulate life through spirits and invocations. The same is alluded to across the Puranas and other related Hindu ancient texts. However, the science of mantras is essentially based on consciousness. It is the *"intention"* that directs a mantra to an *end*, rather than merely the words themselves. By merely appeasing—through invocation—the consciousness-based *shaktis* in creation, the respective mantra is alluded to yielding the *"intended"* result desired by the reciters.

In our discussion about *dharma* we touched lightly on *intention* as part of the dialectical sequence of *"thesis, antithesis, and synthesis"* in action—*karma*. Here again, Hegel's Dialectical Triad explains and clarifies the remote possibility of intention mixed with and transmitted through invocation, especially when in mass, to potentially produce a material influence.

Common mantras said to be powerful in this respect are the *Gayatri* Mantra, the *Kali* Mantra, *Ganesh* Mantras, and the Maha Mantra—each a mimetic representation of cosmic forces.

Gayatri mantra invokes various plenary shaktis, including the *perceivable universe*, the *undercurrent of subatomic, mass-Universal-mind*, the *Sun*, the *Lord in the Heart (Param Atman)*, the *Total Complete Whole*, and the magnanimous *sambandha* relationship between all *that* and you! It is said, with this invocation, you may tap into any desire, although some will argue it's only for spiritual desires. *Intention drives the force of action.*

Kali is the *"dark mother,"* yet is another personification of the Godhead Internal essential Female Principle of procreation, proliferation, and dissemination—*Radha*. Anything you may ask of her is said to manifest.

Ganesh is the demigod of success and the remover of obstacles. Lakshmi is the Goddess of Fortune. There is no shortage of *shaktis*, mantras, and fulfillment of desires to invoke. In the end, the ultimate desire for both liberation and personal *sambandha* with the ultimate Supreme Whole is the *Maha Mantra*.

The takeaway? Mantras are via-medias *(middle ways)* through sound-vibration for the individual to either simply connect with the Universe or create commotion and send out intentions to the various *shaktis*—according to various interpretations.

As to the efficacy of magic in mantras, the Journal of the American Academy of Religion alludes:

> *...mantra is a creative representation and, when uttered, a sonic copy of a cosmic/divine original. ...the perception of this "copy" unites the utterer with the uttered. ...to utter the mantra is to mimic the cosmos/ divine in sound, and to perceive the mantra's sound vibrations is to make sensuous contact with that divine via the "tactility" of the ears and mind.*

> *This is the magic of mantric mimesis [mirroring]. ...This is both magic's biggest weakness and its greatest strength, for while at times it certainly may be "wrong" or unverifiable in the empirical sense, its mimetic logic provides a vitally necessary alternative to the scientific-rationalist perspective, one that emphasizes our connections with the world around us and incorporates, illuminates, and appreciates dimensions of reality and human experience to which the modern perspective is blind.*
>
> ~Journal of the American Academy of Religion Journal Article
> The 'Magical' Language of Mantra
>
> Patton E. Burchett – Vol. 76, No. 4 (Dec., 2008) – Oxford University Press[110]

---

110   Journal of the American Academy of Religion – http://bit.ly/Magical-Language-of-Mantra
        The 'Magical' Language of Mantra – Patton E. Burchett
        Vol. 76, No. 4 (Dec., 2008) – Oxford University Press

We told you to buckle-up!

...

Now consider the exact opposite 50/50—*Yin and Yang*. If positive *sounds, thoughts, and intention*s genuinely affect the individual, communities, and even the planet—through the *sambandha* of *shaktis*—consider the impact of hatred, racism, competition, manipulative religious, political cults, and conspiracy theories. What do we think happens when racial-hatred (*sounds, thoughts, and intentions*) spew from *[bleephole]* QOP clowns as Trump, Jim Jordan, or Marjory Tailor Greene, or the theatrical lies disseminated by religious cult leaders to protect pedophiles and hold on to power and money? Will cults condoning *pedophilia* discourage inequality or domination? Or will Congress coddling and encouraging authoritarian nuclear dictators further perpetuate division or lead to Nuclear War and pursuit of World Domination? We'll just leave you with that equation to solve as the quiz. It's a *"no-brainer"* if we've been paying attention.

Do we now think cults (political & religious) separatists, racists, and conspiracy theories are going away??

Think again!

...

Circling back around and recapping, both Ancient *Puranas'* and Quantum Sciences insist that everything is based on sound/vibration and waves—and mantras are sound/vibration in waves. Did you know that "smell is really listening?" As are sight, touch, feel, fear, anger, hatred, love, endorphins, taste, emotions, and thought? The vibration of molecules is the resonance received and perceived in all our faculties. One explanation of this phenomenon in Quantum Biology is the scent of almonds and cyanide experienced as identical smells. The molecules could not be more different, yet they share the same resonance, and the aromas are indistinguishable.

Back at the primordial fork, when we chose free will and independence, our resolve as independent observers (*ahankara*) elicited the measurable accommodation of ether (*akasha*) and sound (*AUM*) as the foundation for our new temporary finite home. These *shaktis* (energies, plenums, fields), according to both sciences, are *the orchestra and the symphony, and the audience.*

The Ancient Puranas, Quantum Science, and many 'Thinkers' across time allude to a confluence of thought; everything is some kind of *shakti*, particle, field, plenum, or string, including thought, intentions, actions, or feelings—love, hate, fear, hunger, greed, avarice, or even just the ability to feel anxious or tired. The infinitesimal subtlety is challenging to grasp. All these are energies, plenums, fields, in one manner or another, in the primordial soup of existence.

We may not always see the Quantum correlation or the Puranic correlation. So we recap, over and over, to dispel the illusive. Quantum means to measure (from the Latin for "how much"), and *Maya* means "measure" (in its initial Sanskrit definition). Quantum Science alludes that particles are strings, and strings are particles. Both are plenums or "fields,"—loosely—and in the Puranas, *shaktis* are energetic plenums in the foundation of Creationism. Both remind us that indeed "*Measure*" is the ultimate culprit, albeit "Spooky" or "Transcendental," respectively. In his documentary, The Secrets of Quantum Physics, Professor Jim Al-Khalili reminds us: *"No amount of clever jiggery-pokery with our experiment[s] can cheat nature. The two entangled photons' properties [in an experiment] could not have been set from the beginning but are summoned into existence only when we 'MEASURE' them. Something strange is linking them across space, something we can't explain or even imagine, other than by using mathematics. And weirder, photons do ONLY become real when we observe them. ...it truly defies common sense."* – *Professor Jim Al-Khalili C.B.E. - F.R.S. FInstP*

The Puranas describe the living entity (*Jiva-Atman*) as a photon string or pencil of light. *The 24 elemental substances*, described in the Puranas, alluded to particles and plenums of energy that surround, encapsulate, and construct the subtle energetic foundation (*aura*) for material bodies, ability, faculties, senses, species, and thereby the building of ecosystems, cultures, societies, infrastructures, and civilizations. Common sense in both Quantum and Ancient sciences allude to a *cause*, *intention*, or *need* for anything to exist. The Hegelian Triad of *"Thesis, Antithesis, Synthesis"* is the dialectical formula — *"a need, a measure, an intention."* When any particle of consciousness observes a need, a measurement takes place — *what, how much, where, when.* An instance of creation is set into motion, the magic of Consciousness observing the need measures the building blocks required, and the primordial soup stirs into existence required Quanta. It is as if the living spark of Consciousness requests and the *Ocean of Consciousness* churns it into the mix, and a new reality is summoned into existence. This is the allusion projected from both the ancient Puranic and modern Quantum sciences.

Back at our Primordial Fork, as we entered the material expanse, we perceivably reduced the manifest surrounding vibratory commotion into visual-auditory interpretations—of that elusive *measurement of the unmeasurable—Maya or Quanta.*

How does the Ultimate Observer measure the unmeasurable? Reductionism. The new material expanse insists on reducing everything that is infinite into finite units of measure to accommodate the limited perceptual illusion. Whether we, the *Jiva-Atmans*, are photons of light or Electrons, Bosons, or Quarks is insubstantial and just as unmeasurably unimportant—to the infinite degree. Such comparison would be mere semantics. What matters, ultimately, is the similarity in "*Mechanics*." Both ancient and modern describe particles and waves from energetic and vibratory sources, playing a game of *Hide-and-go-Seek*—creating phenomenological universes from the inside out, the bottom up, from an internal subtlety of dialectical measure and intention creations blossom like flowers in a field. We, the observer(s), are the seeds. The perceived needs, by the observer, dialectically manifest particles of intention that were not there before ideation for any action or journey.

Over time, 'thinkers,' philosophers, and scientists repeatedly studied the Ancient Texts and deduced similar subatomic findings to modern Quantum observations. In the end, both address the soup-like compound solution of substances and plenums intermingling and permeating the manifest universes. "Substances," "plenums," "fields," "waves," and in the Puranas, an "*Ocean of*

Causality and Consciousness," seem to be the consensus among thinkers, philosophers, and scientists.

You May Skip the Rest of this Chapter if New to Mantras or Hinduism

Then there is the first line of Gayatri Mantra:
AUM bhur bhuvaḥ svaḥ 'TAT'

Coming back full circle, the power of chanting the syllable *AUM* is that by itself it is ecumenical, all inclusive, without borders or distinction. *AUM* encompasses not only the meditation and invocation of all forms and *shaktis* from *Vishnu*, *Krishna*, and *Brahmán*, but it does so without distinction, difference, or separateness. In fact, *AUM* is so all inclusive and universal that it is found as the seed mantra invocation in other traditions of spirituality, including Tibetan Buddhism, Sikhism, Jainism, and all forms of Hinduism. In essence, *AUM* is just universal and without boundaries or judgment.

The above *Gayatri* Mantra, for example, *AUM bhur bhuvaḥ svaḥ TAT*, creates an eye-opening meditation that begins to revel deep within the *sambandha* universal relationships of plenums (*shaktis*)—from our gross matter reality to the subtle existence of thoughts and intentions. The Puranas explain these *planes* as essential foundations of our very existence. The *Gayatri mantra* perspective delivers a cascade of the primary essential three primary plenums, according to the Puranas, of our sub-atomic levels of reality.

So far, we have discussed only some of the many meanings of the sacred syllable *AUM.* The next Sanskrit word, *bhur*, implies *"our plane of existence, earthly, worldly, material, which we experience."* However, *bhur* also means *"prana, life, or life-breath"*—combined, the material composite manifestation of life-force and consciousness—the visible material world.

The third word *bhuvaha*, invokes the next plane, the all-pervading mind consciousness expanse, ubiquitous throughout all celestial space (*akasha* or ether). This indicates the subatomic layer, where energies from the many vibrations and wavelengths interact and dance altogether beyond our perception—the universal, all-encompassing connected *ubiquitous-mind* state theorized by Descartes and others.

The fourth word, *svaha*, takes us even deeper down that rabbit hole— beyond what we call the sub-atomic and *ubiquitous-mind* layers of existence—the primordial *24 substances* and innumerable *shaktis*. This expanse is the 'other side of matter', the non-material world of only pure consciousness—the spiritual world. This *svaha* is the all-pervading nature and omnipresence that pervades the multiverses—the Persona that is the source of all manifest creation—*Vishnu.*

The last word in this first line of the *Gayatri Mantra* is simply the powerful "*tat,*" literally meaning "That." In context, *tat* implicitly signifies "The All," "the Complete," "The Cause of All Causes," "The Supreme Consciousness," "The Tao," "The Godhead," "the Supreme Persona!" This word tat is the aggregate of the above triad (*bhur, bhuvaha, svaha*) in relationship to the self and the world—altogether *AUM.*

And this is only the simple explanation of that first line from the *Gayatri Mantra.* In future volumes, we will dive deeper into this mantra meditation's second line, which deals with the *sambandha* relationship through the mind and mindfulness meditation. Before we continue down the rabbit hole, however, we need to revisit how we arrived at this explanation of *AUM* and the beginning of the *Gayatri Mantra*—understanding the Puranic Quantum science of subatomic substances.

The *Puranas*, *Vedas*, and *Upanishads* allude to *Bhakti Yoga* being the ultimate goal and highest mindfulness, but not until the Quantum-like Science of Sambandha is "Realized." Without truly, sincerely, and profoundly perceiving the subatomic *shaktis* all-around, *Bhakti* remains merely sentimental religion. Modern scholars, teachers, and gurus indicate that simply skim-reading about *Sambandha* and then establishing a temple is the end-all— and then *"just give us your money so we can make millionaire swamis even richer."*

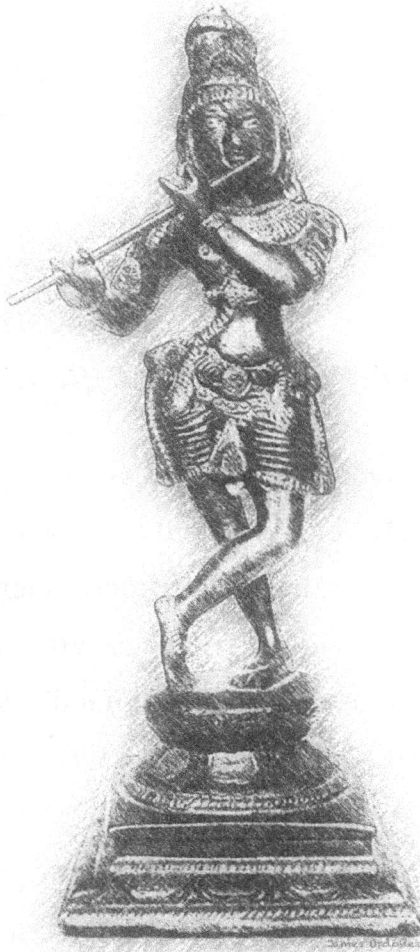

Krishna - The flute player Cause of All Causes
(Illustration by the author, James Ordonez from personal photos)

Chapter 9

Why Bhakti Yoga

Literally, the *"Yoga of Devotion"* is said by some to be the highest form of Yoga, but only when certain conditions exist, according to the Puranas. The practitioner, spiritual group, or preceptor first chooses a deity from a carving, painting, or other traditional objects representing one of God's various forms in Hinduism. Traditionally, some rituals and *mantras* recommended summon the presence of God into the article. The practitioner then daily dedicates offering prayers and offerings of love as unto God. The practice becomes a lifestyle, and a holistic meditation ensues.

Beautiful, ...No?

Okay, let's get one thing straight, *Bhakti Yoga* is not complete without everything we have been exploring from the Puranas. *Bhakti Yoga* is the practice of personal care and service to the

Gods in full meditation, with food and drink offerings, pleasure and comforts, reverence, and devotion—through the deity. This practice includes song and dance for the deity's joy and mantra meditation, which the practitioners take with them day in and day out. This practice is a sublime, holistic 24/7 domestic meditation that engulfs the practitioner's life in awareness of God at all times—if done undeviatingly and with the correct understanding of *sambandha* and all its principles.

The challenge is, without a full grasp of all the before-mentioned Puranic science of *sambandha*, the *24 substances*, all the *shaktis*, *dharma*, and the complete comprehension of *AUM*—all together—the practice falls flat. The meditation becomes and remains a mere sentimental practice of idolatry, not understanding what one is focusing upon. Of course, there is some value and advancement from any sincere intention. Yet, the shallow meditation without the deeper Puranic Science of *sambandha* precludes the practitioner from achieving the following two stages of realization explained in the Puranas. Until now, we have only discussed *sambandha* with all the essential components. While we will not go deep into the next two stages in this edition, some brief explanation and understanding are required to comprehend the relationship of *sambandha* to *Bhakti Yoga*.

> *Without a solid grasp of Sambandha all one's bhajan (spiritual attempts) remain on the neophyte level."*
> ~*Guru Bhaktivinode*

Abhidheya is the practicing stage after *sambandha*. This stage comprises the functional transformation where the practitioner employs *karma* (action) with *universal understanding, profound comprehension, and perpetual immersion* in *sambandha*. Then, *Bhakti Yoga* starts the holistic journey to achieving the goal of the Puranas. The last stage, called *prayojana*, means the ultimate goal, combining the previous two phases of development. We here circle back around; to the dialectical journey of the *jiva-atma—thesis, antithesis, synthesis* —the dialectical triad discussed in previous chapters.

> *Use Common Sense, and if you have none, consult with someone who does!"*
> ~*Bhaktivedanta Swami*

Genuinely understanding the essential science of *sambandha* from the Puranas, we may agree through *Common Sense* that this "Science" is the quintessential most crucial component in the equation. Without *sambandha*, there exists no true *bhakti*—only sentimental idolatry. We have mentioned before the quote from *Bhaktivedanta Swami*, "Science without religion is mental speculation and religion without science is sentimentality." In this context, *Swami Bhaktivedanta*, the original Guru and preceptor of *Vaishnavism* in the West, would often state, "You must judge *[a thing, process, or action]* by its results." Indeed, the incessant fighting among "the cult" members and splinter groups, their criminal activity, lack of human empathy, and child abuse tolerance in the cult do not indicate any devotional result. Therefore, we wrote this book to implore any who wish to become acquainted with and even practice these ancient sciences to do themselves a favor and remain wholly detached from "the cults," especially this one. The mere association, *karmically*, with those tolerating such misdeeds, will—according to the Puranas—deter and even block any practitioner from any actual spiritual attainment. This instruction is among the conclusions of the Puranas.

There do exist some *Vaishnava* groups *(sanghas)* that may be followed safely, of course. Such are those *sanghas* free of these vices and corruption by mere dint that they are not "organized" with money, Real Estate, politics, and oligarchies. These organizations are good to visit, albeit detached, for inspiration, classes, and congregational *mantra meditation (kirtan)*. We will name a few. It is often best to regularly visit the nearby traditional Hindu temple and stay away from Western gurus and Western groups. In any case, a good rule to use, as a barometer, is to consider visiting the groups or gurus demonized or shunned by "The Cult Branch."

There are mainly two kinds of groups to seek out, those within the *Gaudiya Vaishnava Lineage* and those independent. If available in your area, some *Gaudiya Vaishnava sanghas* still free of avarice, and denominational politics are The *SCS-Math* (B.R. Sridhar Swami Sangha)[111] or the *Narayan Maharaja Sangha* (Also known as Pure Bhakti).[112]

☞ *The careful consideration to watch out for is a region by region festivity overlaps with "The Cult Branch." When this mixing happens, a danger exists of getting sucked in unconsciously into "The Cult" because it is more extensive and politically more powerful and influential. Again, "cults" are insidious and addictive by their nature!*

We continue to explain this subliminal danger in associating with cults—pervasively alluring and indoctrinating. Be extremely careful of the larger establishments. The other type of *sanghas* are independent of the *Gaudiya* lineage, yet still hold dear many Puranic essential practices of *Bhakti-like* congregational *kirtan*

---

111   *Sri Chaitanya Saraswat Math - http://www.scsmath.com*
       http://www.scsmath.com
112   *Narayan Maharaja's Pure Bhakti - https://www.purebhakti.com*
       https://www.purebhakti.com

and discussions from the Bhagavad Gita and other Puranas. One of these is the *Amma Ji Sangha* (aka Amritananda Yoga Ashram).

> *Every endeavor is covered by fault, as fire is covered by smoke"*
> ~Bhagavad Gita Purana 18-48

Some, not all, of Amma's followers believe and proselytize that she is an "Avatar" (God descended to earth in human form). Of course, when we discussed the Puranic *sambandha* science of "simultaneously one and different" *(achintya bheda abheda)*, we learned the Puranic proposal that everyone is a spark of God in human form—simultaneously. The danger of this differentiation— that only one living being reigns over all others as an avatar from God—is the seed planting for a cult to begin forming around the precept. Still, Amma's group's *kirtans*, gatherings, and classes are the most ecumenical and least sectarian. Amma never asks anyone to change their religion; instead only encourages all-around love for the world, others, and oneself. So, by all means, visit, stay a little while but don't get sucked into believing she is God. She will never say so; she will only wish to hug you and whisper a loving mantra in your ear, the Sanskrit letter *Ma* repeatedly, the third letter from the *AUM (Om)* chant *(also meaning mother)*. She is therefore known as the hugging Guru.

"The cult" in question, on the other hand, explicitly teaches that "the one and only way" for one-and-all is to surrender to "them" (the leaders) implicitly and no other group or even slightly differing belief. Worst of all, if one criticizes or offends any of them or their clergy for tolerating or committing child sexual abuse—or other criminal activity, including murder—then "one's spiritual life is doomed forever." The "offenders" then return to the material worlds always in hellish existences. For all eternity, they are doomed until forgiven by those who have been criticized! They call this *"aparadha,"* a Sanskrit word literally translating to "anti-the-Goddess" *(Radha)*. Those who contradict or fault "the servants of God" will suffer the very worst suffering in creation. Sound familiar? At least in the Catholic Church, they just ignore critique; in this cult, one suffers eternal damnation for *"finding fault with the clergy."* So we implore you yet once again, avoid these *[bleeping]* knuckleheads and all their members to ensure you do not get inadvertently sucked in. It seems to always happen to the least likely—getting caught in the *spider web*, that is.

The other two groups have similar beliefs about *"aparadha,"* however, they do not use this precept for political control over their members and followers. Both groups are splinter offshoots of the original Hare Krishna import in 1965 by Bhaktivedanta Swami, after his passing when many of his original mainstream disciples went criminal and RICO[113] *(Racketeering Influenced Corrupt Organization)*.

Nevertheless, the parallels between political cults and religious cults speak to humankind's avarice nature *(acquired dharma)*. A cult is a cult with all the ingredients of a cult—political or religious. When one compares the clowns in politics to the clowns leading religious cults, the only difference is the dress. A cult is a cult, political or religious. Therefore, just stay away from all *"Mr. Swami's"* toxicity and their supporters! Detach for your own material and spiritual health and that of all those around you.

Sadly, like any cult, most of this *"Bhakti"* Cult's followers are kind and innocent folks. They are guilty only of being brainwashed and indoctrinated, rendered slaphappy to tolerate indiscretions. Yet, thereby they are complicit in inadvertently supporting. Such is the making of any kind of cult, as discussed earlier. Isn't that the case with the millions who follow clown politicians not understanding the rhetoric being fed to them is complete *[Bull-$#¡t]*? The tribal herding instinct for community and belonging is as strong as sex drive, as addictive as drugs, and is easily manipulated by cult leaders quite consciously and effectively.

The entire premise of *Bhakti Yoga* becomes compromised. Tainted

---

113   *Encyclopedia Britannica – http://bit.ly/RICOorgnaization*
       R.I.C.O.

with *group karma*—despite the misinterpretations that *karma* does not apply to the cult because "they are engaged in *[so-called]* Bhakti Yoga." That ends up being a vicious catch-22 cycle, where indiscretions begin to grow in all directions, the more the belief of immunity saturates the members. Every kind of cult has some form of this justification, believing deeply in their hearts that their mission's righteousness precludes consequences—naturally. It's a madness brought on by the dialectical illusion of the same *Maya* or devil these cults crusade or campaign against.

☞ *Three other recommended books for seekers interested in 'carefully' choosing a Bhakti Yoga community are 'MONKEY ON A STICK'*[114] *from the late 1980s, Nori Muster's BETRAYAL OF THE SPIRIT*[115] *in 1997, and Henry Doktorski's book series, the 2018 KILLING FOR KRISHNA.*[116] *All three exposés give many witness accounts on massive corruption detailing the cult's criminal activity over the decades.*

Nothing much has changed, only some players and some creative *market-positioning and posturing*, just as it happens with political cults. "*The cult*" empire has grown exponentially with Multi-Millionaire Swamis, gargantuan buildings, Real Estate, ongoing scandals, embezzlement and laundering, and regular reports

---

114   Monkey on a Stick - http://bit.ly/Monkey-on-a-Stick
       TIME Magazine Review: John Hubner & Lindsey Gruson's Monkey on a Stick
115   Betrayal of the Spirit - http://bit.ly/Betrayal-of-the-Spirit
       Amazon Books Nori Muster – Betrayal of the Spirit
116   Killing for Krishna - http://bit.ly/KillingForKrishna
       Amazon Books - Henry Doktorski's Killing for Krishna

of justified reckless and dangerous criminal activity—always somehow for the mission. It's telling how money has a way of sweeping many things under the proverbial rug—as with political cults. And like all cults, the unknowing "innocent" remain culpable for tolerating and continuing to give support through their presence, service, and donations. QAnon or "The Cult", they are all driven *[bleeping]* insane. One has to ask, "Does '*Bhakti*' truly exist in such environment, and to what degree, especially without "a solid grasp of [the] *Sambandha*" science from the *Puranas*?"

...moving on

Both Narayan Maharaja's Pure Bhakti group and the Sridhar Vaishnava group splintered from "The Cult Branch" in 1977-1980 due to the criminal activity and have remained ethical and law-abiding ever since. They have their issues and *characters* but are not constantly being scrutinized by law enforcement. We personally stand behind and support them both. That is not to say there are no issues. Of course, there are; but overall, they are free of corruption. However, as with the *Amma Sangha*, it remains unclear if they embody the full grasp of *sambandha* science as described throughout this book from the Puranas—what to do! *(Give them all this book?)* We find this elemental foundation of *sambandha* is typically upstaged by the sentimentality of religion alone in groups; therefore, the individual needs to know and be proactive in developing their personal *adhikar* relationship with the Puranic science of *sambandha*.

With all the above, probably the best manner of performing *Bhakti Yoga* is at one's own home, with only inspiration from a community or group—removed at a distance. Deep familiarity with a group often becomes a sort of a spider's web, *"Welcome into my parlor," said the spider to the fly.* In India, we find that most families and individuals manage to maintain their various spiritual and religious practice at home while only occasionally visiting a temple for services, weddings, funerals, and religious holidays. Ashrams are typically temporary retreats or shelters for retirees, the needy, or schooling years.

According to the Puranas, *Bhakti Yoga* can be the highest Yoga and should be pursued as an end to all ends. However, without complete understanding and meditation on *sambandha*, it remains lower than *Karma yoga*, which at least focuses on helping and serving others as equal parts-and-parcels of God. This logic arguably positions *sambandha* above even *Bhakti Yoga* as the essential prerequisite and ongoing necessity to continue.

As with *Amma Ji*, the hugging saint, and her followers, "love," for everyone and all, alone can be converted into *Bhakti Yoga*—once the *sambandha* realization is securely embedded in one's *dharma*.

One must first be convinced beyond the shadow of a doubt that every speck of dust in the universe and every living being are simultaneously one and different with *Vishnu*, *Krishna*, God— without exception—only *infinitesimal*. At that point, one's "*Love*" itself also becomes God—*simultaneously one and different*. But only until that very lofty distant goal. In the meantime, aspire but continue meditating on *sambandha* with *mantras* and *kirtan* and *Karma Yoga*.

"The cult" will try to tell you differently, but this is the allusion from the Puranas—without dependency on a tribe or cult to tell you how to think.

By studying and meditating on the Puranic Science of *sambandha*, the seeker or student slowly becomes aware of their relationship to the universe, the cosmos, the spiritual and material substances of creation, God, and all other living beings. With these meditations, mindfulness, and regular practice, the self develops the internal peace to see God's presence in every being and every object. That is when *Bhakti Yoga* begins to blossom from within—according to the *Puarans*.

Then, the offering of flowers, incense, or service are no longer limited to the deity on the altar, but in one's heart, to everyone and everything around you with empathy and compassion, the Love of knowing that everything is a spark of the same God. That is *Bhakti Yoga*! Without *sambandha*, there is no *Bhakti*.

The alluded conclusion on *Bhakti Yoga* is that without the Quantum-like science of *sambandha*, the implication is religious sentimentalities, which drive idolatry. Beliefs on a host of shaktis (sometimes known as Demigods) within the material expanse selectively become little more than idols. The Puranic "Ocean of Consciousness," known as "The Cause of All Causes," also become sentimental idols *without* this Relativity Science from the Puranas. The reverence goes out the window, misplaced. The logic enunciated by figures like Bhaktivinode Thakur follows. "Sentiment" for a host of "divine energies" is merely a neophyte necessity. Without the Relativity Science of *Sambandha*, are born tribalism, sectarianism, posturing, and (religious racism) separatism. They invade the rhetorical dialectics of religiosity—making total hypocritical asses out of cult leaders and their followers. The aggregate allusion from the Puranas is to first embrace *Karma yoga* and *Sambandha Science* as a means to develop and understand Bhakti without mere sentimentality.

Ganesh - Shakti of success and remover of obstacles
(Illustration by the author, James Ordonez from personal photos)

Chapter 10

Dangers with [bleeping] Cults & Organized Religions

a few short, eye-opening true stories to illustrate a point

Important Note:
These stories are only from the "The Cult" discussed, led explicitly by "Mr. Swami," NOT about any other Vaishnava or Hare Krishna religious group.

These are personal recollections. My first scare, which like a *[bleeping]* idiot, I did not heed to the warning, was my first departure from the cult. I had gone to spend the summer *with nature* at the West Virginia farm. I received a bunk in a cabin to share with another member and three square meals a day in exchange for *service*. I would wake up before sunset and spend the days taking care of and working two large beautiful brown and white stallions, *Tom and George*. The latter was the mare, and they would not go anywhere without each other. After brushing and feeding them each morning right after breakfast, the three of us would spend the entire day dragging and shoveling large wagons of lime across several fields for next year's harvests. It was a beautiful summer until it was time to get back to city life in the fall.

I approached the leader, Bhaktipada,[117] to let him know it was my time to depart after a fulfilling summer with nature and the horses. I even suggested I would be coming back the following summer. To my dismay, his reply was abrupt, "*You're not going anywhere!*" He turned to our posted-child, "RatNut Swami," and gave him a clear look of, "*you know what to do.*" RatNut was not yet a "Swami," he was merely known as RatNut Das. I had heard rumors of dead bodies buried in the hills; now, I was a little concerned. I left the room thinking, "I have to get the *[bleep]* out of here!!" I had to plan an escape. I harnessed the horses and went out on the fields as if not at all concerned. It began. Everywhere I went across the fields or daily functions, I would see "RatNut" standing in the distance, his hand on his chanting-beads, simply staring in my direction, with two others

---

117   *Killing for Krishna by* – *https://bit.ly/Killing-For-Krishna*
       https://bit.ly/Killing-For-Krishna

by his side, day and night. This went on for days. A week and a half passed, it was 1:30 am; I quietly rolled up my sleeping bag and grabbed a backpack, and slithered out towards the back end of the property to avoid the main entrances. Flashlight in hand, I hiked through the brush and streams in a straight direction until dawn lit up the sky. There was a small country road, and I began to hitch-hike back to NYC. That was my first escape.

Fast forward...

It was the late 70's. I took a job in Los Angeles as Production Manager at F.A.T.E. Studios, a small museum sculpting company constructing and designing dioramas for a new Hare Krishna Center museum project. One day my partner and I walk out of a building on Watseka Avenue in West Los Angeles, across from the Hare Krishna temple. A sight I can never forget and often find difficult to even forgive. The Krishna Temple's head cook at the time, Sruta, was violently wailing a solid oak, meter-long ax handle at a young man's head and torso on the ground. Shirtless, wearing a white Indian traditional dhoti from the waist down, Sruta had a shaved head and small ponytail dangling from his head's crown. A somewhat corpulent monk, *Achyuta-Something Swami*, at his side circling, screaming, "beat him 'till he barely breathes, that's right, 'till he barely breathes," hollering over and over, and over. After several dozen blows, the young man, Dave, down on the concrete sidewalk screaming in pain, manages to get up and run staggeringly to the corner of Watseka Avenue and Venice Boulevard, Sruta chasing him with the ax handle. It was five in the afternoon. The sun was burning down on a late summer afternoon with heavy traffic on the six-lane boulevard. Cars were rubbernecking in horror at the sight of Sruta once again, continuing to beat Dave incessantly without mercy. Again, Dave was able to get up and ran into the perilous traffic, where Sruta could not continue the chase. Almost being hit by several cars, Dave ran into Balians, a supermarket across the street and down the block. Sruta then calmly walked back up Watseka Avenue with ax handle in hand, conferred with the Swami, with a pat and a hug, and went to the back of the temple building where the kitchen awaited him to finish cooking that day's religious holiday feast. I don't recall what the Holiday was, only that Dave had come to help in the kitchen.

Cell phones, of course, had not yet been invented. When asking around for a landline from the cult members, I was told several calls had already been made, but to accuse Dave rather than report his attacker. No one ever came, no police, no ambulance. After some time waiting for the police, I walked across to Balian's Market. I was told they called and handled the police and ambulance. Paramedics had told them it appeared that every bone in his upper body was broken, and most of his ribs, including arms and concussions to the head. "It was a miracle he survived," said one of the cashiers.

Dave was a regular guest at the Sunday Feasts, a traditional weekly proselytizing free meal gathering, and other religious holidays. Dave seemed around 19 or 20, a handsome young man with a lisp, had reportedly been coming around for a couple of years, volunteering to wash pots and cut vegetables on Sundays and holidays.

Still expecting the police to come around, I began to inquire about the incident. Sruta, the cook with the ax handle, told me that Dave "had raped a *Brahmacharini*," the Sanskrit word used for female *Brahmachari, temple devotee*. A *Brahmachari* is a celibate male monk. The celibate monk status was traditionally reserved only for male monastery ashram students in India. When Bhaktivedanta Swami came to the West, he allowed women to join the ashram. Swami Bhaktivedanta understood that breaking the archaic patriarchy was needed and the tradition to be left both in the past and in India. He was somewhat progressive and was criticized for this by his peers for this and other decisions regarding equal rights for the female members. Nevertheless, in my inquiries, I found out that Sruta was not at all accurate in his rationalization for the assault.

On the one hand, the girl in question was not a *"Brahmacharini" temple devotee monk* living in the *ashram,* but rather another regular Sunday guest, reportedly dating Dave for several months. They had been seen coming and leaving together on several occasions. According to some temple members, she and Dave went to the beach on Sunday nights after the Feast to be intimate. Others said he was forcing her to leave the temple with him every Sunday night, for months, to go five miles to the beach to have nonconsensual sex. It was apparently an ongoing saga among the temple gossip. She allegedly was approached by the women in the ashram, preaching to her on the evils of sex. The rumors were of repeatedly having "illicit sex" on the beach after coming to the temple every Sunday. This was deemed "demoniac" for using the temple gathering as a venue to pick up women." It was all hearsay and gossip, but there was enough clamor on several sides about the ongoing weekly beach sexual encounters to give it a history. Apparently, they were, in fact, dating.

☞ *Note here the dialectical creation of a reality for an entire community based on dogma, taboos and untruths.*

This faction of the Hare Krishna sect secretly teaches that sex outside of marriage is demoniac, therefore taboo. When I pressed, was told, she alluded to having been seduced by Dave. The news ran out like a wildfire that she had been pressured to have repeated sex on the beach Sunday after Sunday with Dave. By the time the gossip reached the kitchen, Dave had arrived for his weekly service. It never became clear exactly when the seduction story turned into "pressured," which became "forced rape," but that was their interpretation.

Sruta was known for having an angry streak. Still, this was beyond the pale for a *[bleeping]* monk in robes with a "Swami" circling like a cheerleader. Someone had used the words to Sruta, "Dave raped a Bramacharini," even though all the facts were very questionable. She was clearly not a temple ashram devotee, and they had apparently been dating. Sruta, catapulted by the cult's taboos on sex, the ugly rumors handed to him by word of mouth, his own anger issues, and the circling corpulent Swami, took the law into his own hands. He beat this young man practically to death—in full rush-hour public view, with a "Swami" egging him on and with no remorse whatsoever.

Everyone just stood around watching in horror. A couple of us run over to try and stop it but could not get close, the meter-long ax handle wailing furiously. It was then when Dave found an instance and strength to get up and run.

Reportedly, neither Dave nor the young lady ever came back. I could not find out if it was related to this crime; however, I learned years later that Sruta did five years in prison and later passed away.

• • •

With these types of criminal behaviors popping up across the cult's global communities, many of us just left altogether without looking back.

In the mid-nineteen-nineties, a few independent News Websites popped up, which various Vaishnava groups used for community reporting and article submissions from members. One was called VNN.org (Vaishnava Network News), another Chakra.org, and another still around called the Sampradaya Sun (sampradaya.org). Many of us had left but remained curious about the health and safety of those left behind, any new politics, and what became of friends and acquaintances. Articles began popping up every-where about a murder in the early eighties. A young man named Sulochan Dasa, from the farm where "Mr. Swami" lived and helped manage had reportedly been murdered through an alleged conspiracy that to this day implicates "Mr. Swami"—our poster child. Reportedly, "Mr. Swami" was given a plea-bargain deal by the FBI to *"rat out"* his accomplices by the prosecutors, which, in turn, sent several others to years in prison for conspiracy to commit murder. To this day, *"Mr. Swami's"* disciples, followers, and staunch "Cult" followers claim it was all a lie "manufactured" to implicate "Mr. Swami." The readers can judge for themselves with dozens of accounts.[118]

The internal fighting in the cult continues to this day over murders and child abuse. And guess who is still at the center of all of this over and again? You guessed it, *"Mr. Swami."*[119] [120] [121] [122] [123] [124]

☞ *Interestingly, a recent defense within the cult is, "The ISKCON Governing Body has determined by their "own" investigation that RatNut Swami is not guilty of murder conspiracy because allegations reported everywhere are incorrect." Isn't that like the Vatican declaring for centuries that alleged child abuse was a fabrication, or the QAnon cult preaching that Trump never "grabbed pussies," never told lies, and remained President after he was voted out? A cult is a cult!*

118   Murder of Sulochan Das conspiracy – dozens of search results – http://bit.ly/MurderOfSulocan
       Murder of Sulochan Das conspiracy search results
119   ISKCON Truth Stories Website Site
120   Killing for Krishna – By Henry Doktorski – http://bit.ly/KillingForKrishna
121   Book: Killing for Krishna by Henry Doktorski
122   The Sampradaya Sun – http://bit.ly/RadhanathAndTheKing
       News Website: When Kirtanananda was king and
       Radhanath's involvement in Sulocan's murder
123   Birds of a Feather – http://bit.ly/LA-Times-Report
       L.A. Times Report — When Radhanath Swami was the right hand man to Bhaktipada
124   Beyond The Los Angeles Times Report – http://bit.ly/BeyondTheLosAngelesTimesReport
       Editor's Clarification

[WTF]! With so many allegations throughout *Mr. Swami's* career, it was very much like Trump's statement regarding Putin, *"He said he absolutely did not meddle in our election; [therefore] he did not do what they are saying he did."* –Donald J. Trump.[125] Do you remember that *[Bull-$#¡t]*, egging on and encouraging Mr. Putin to meddle with the World? "To tolerate is to encourage," much like those who protect *Mr. Swami—a cult by any other name.*

Nevertheless, in my travels, I ran into the Krishna farm in Mississippi, where I learned from a very reliable source that *"Mr. Swami regularly still visits."* He comes to see one of the men he ratted out, who allegedly went to prison for him, and each time hands him *"Wads of cash, a lot of it, at every visit, a few times a year."* Draw your own conclusions there.

These stories and many like them are some of the reasons for this book's writing—how sinister a cult's misguided beliefs can manifest into criminal behavior covered up by the tribe's comradery. Moreover, these stories also give light to the total disregard for the ancient texts and science they claim to represent, as is the case with most "organized religions" and tribal cults.

---

125   *Trump Believes Putin – http://bit.ly/TrumpBelievesPutin*
       NPR Report Trump Believes Putin Despite Many Intelligence Reports

...

After a couple of decades, I decided to check in to see how things were going. It was 2003, and I brought a few people together. We lined up retirement pensions and Social Security resources for the members. We started a small Hospice in India for the group, funded by a one to five dollar PayPal donation global campaign. We copied other churches' models for member care and attempted to sell the cult leaders the ideas of Social Security, Group Healthcare, and Retirement Fund accounts for their aging members. We started a web presence, created audio visual Power-Point presentations, printed brochures and marketing materials, and began gathering interests from the group members worldwide. Not all were happy with our "meddling and trouble-making." Some of the leaders were furious and became outright hostile.

...because that was not [bleeping] telling!

In any case, this networking for Social Services to the community did open some doors, offering our cause an opportunity to be heard seriously. Subsequently, I was hired to act as the group's legal secretary to a Court Ruling Monetary Fund Dispersal of twelve million dollars awarded to ninety-five claimants of a child abuse case, known as the Turley Case. The dispersal group consisted of a dozen and a half Temple Presidents and Board Members of the cult's governing body. The accused had been elementary school teachers, Swamis, leaders, and gurus of the 1980's "Cult Branch".

The beneficiaries, victims, were now in their early twenties, going on with their lives, albeit scarred and dysfunctional to one degree or another, most having to continue ongoing therapy in order to cope. My job was to take objective notes and report from a series of meetings led by a businessman, congregation member, named to lead the strategy. The courts had requested that "The Cult Branch" group manage and propose a strategy back to the courts to distribute the twelve-million dollars awarded funds to ninety-five victims—*approximately six figures each*. There were complications. A number of victims suffered much more than others. Some of the details were so gruesome that it became challenging to stay objective. A few of the children had only suffered one or two sexual assault incidents, while (on the extreme) one had gone into convulsions at five years old and hospitalized from having "several cups of semen removed from his large intestine." The courts had allegedly agreed that the burden of a fair distribution needed to be managed by the organization responsible—"The Cult Branch." Frankly, I never understood how a court would give such a responsibility to the perpetrators, but I was only being paid for writing and reporting each meeting's proceedings objectively.

Despite the gruesome details, I was still able to remain objective until something just as unthinkable happened. Representing the New York chapter, *Ram B*, a "Temple President," stood up in defiance of the victims seeking compensation. "We have to punish them for stealing from our Guru's mission," he declared over and again in so many words. His recommended solution was to ensure that they "do not receive their due compensation by increasing the number of claimants." The original lausuit filed was $400 Million, and had already been reduced by the Court to $12 Million. The logic he argued was that if equally distributed, each victim would receive around $100,000. On the other hand, *"if we get ten thousand claimants, each will only receive a little over $1,000."* The room went wild, with half the group calling it genius, cheering in excitement. I and the other half of those present just got up and left the room. Some were yelling and screaming on both sides. It was a clear split.

> *We have to punish them for stealing from our Guru's mission, ...if we get ten thousand claimants, each will only receive a little over $1,000."*
> ~Ram B.

In the end, the vote was to proceed with the "punishment" as conspired by those SSPT *[bleephole]* members voting to add salt to unspeakable wounds. I was not a voting member, and I know that a couple abstained. Another few voted against such insufferable retribution to innocent children seeking compensation from atrocious acts against them. Only a couple of votes won the horrific decision, and a global campaign went out requesting anyone who had been even as much as yelled at or chastised by their school teachers or cult leaders to come forward and collect a piece of the pie. Ironically, the acronym SSPT stood for "Spiritual Strategic Planning Team."

☞ *"The religious order placed advertisements in newspapers and other media seeking alleged victims of abuse. More than 430 came forward in addition to the 95 already named in the lawsuit."*[126]

~ Los Angeles Times

It worked, but not quite ten thousand came forward. As I recall having left again in disgust, the last number I remember hearing was that a couple of thousand came forward, yet with ludicrous accusations of someone grabbing their arm or raising their voice. The ad campaign included personal coaching and witness leading of what and how to say it. Among the campaign leaders were the temple presidents from New York, Los Angeles, Dallas, and other Board Members of the cult's Governing Body. The court dismissed many as frivolous but the number still climbed up to

---

126   *Turley Child Abuse Case - http://bit.ly/Turley-Child-Abuse*
       Los Angeles Times

[reportedly] disputed 535 claimants, that if evenly distributed would have been less than $20k.

Somehow, this truly despicable and evil punishment of the victims went well publicized through the group and was, in the end, partly approved by the Courts. Folks came forward for all types of complaints, that were not really "child abuse," some even only bordering on "they were mean to me." Of course, it's always "the mission first, at any cost to others," as mentioned in earlier chapters.

☞ *Note here again the dialectical creation of a reality based on dogma, taboos, and untruths. The deception was "positioned" in the proposals to the courts and all cult members as an honorable attempt to compensate "everyone" who had ever been abused. Even the judge and opposing attorneys bought the [Bull-$#¡t]! A "reality" was created dialectically, which was, in fact, total lies. Subsequently, all in the SSPT became heroes even in the eyes of the Court and opposing counsel.*

Among the folks who voted for the appalling notion at the meetings and many others who willfully joined the *punishment campaign* had previously been folks of character, conscience, and empathy. The cult had taken that away from them. It was as if they had been stripped of any decency and civility for the sake of this ungodly agenda. This, for me, became both unforgettable and unforgivable, a monstrous act of cowardice and vengeance against innocent children. *(I am certain the readers may agree with the sentiment that these are nothing but [bleeping] monstrous [bleep-holes], not spiritual people!)*

Reportedly, one of the leading child abuse perpetrators over the decades is "His Holiness," *BhavaSomething Swami*. Some rank and file child abusers get excommunicated when the law gets involved in Western countries. Others of stature or connections remain protected by the cult leaders. The most revered cult leader, "His Holiness," Mr. Swami, or any of his Cronies, have always had the power to remove him any time since *(now over four decades)*, but guess what? *BhavaSomething Swami* was not only a Swami but also a "Guru" and a member of the Governing Board Commision. Therefore, his punishment for proven repeated sexual child abuse was a demotion from the above three positions to merely a "teacher." *(You can't make this [$#¡†] up.)* These cults very often turn to monsters in the name of some "mission for God." The Internet is riddled with this cult's Pedophilic accounts.[127]

To this day, these types of behaviors by the group's leadership are quietly swept under the rug by all cult members. After all,

---

127   *Child Molesters: Gurus? – http://bit.ly/Child-Molesters-Gurus*
        "Child Molesters: Gurus?"

once a person joins a cult such as this, both their success within the group and even their livelihood are impacted if the cult does not succeed. An ugly, and yes, arguably "demoniac" symbiotic relationship develops, and everyone, to one degree or another, becomes complicit, in the name of God—*"The Company First!"*

...

Fast forward to 2013. Ten years after leaving again for the above horrific set of circumstances, our hospice in India was still running. Still small, but PayPal donations were still maintaining it, and a small handful of devotees had nursed it along all those years without my involvement. Clearly, there were still some good folks holding on to empathy and goodness. That year, passing by, I visited one of the cult's farms in Pennsylvania named Gita Nagari. Finding out that the farm was not under *Mr. Swami*'s control gave me the reason to attempt another business plan for a retirement village for the *uncared-for,* aging members, with all the retirement resources we had been peddling.

Again, we formed a Board of Directors, lined up Government Farm Grants for housing and small business loans, and the retirement resources discussed. We received approval and support from the farm's local leaders and their guru in charge— Devamrita Swami (one of the "nice guys"). Everything was set to go. Professionally, we, of course, mentioned it to "Mr. Swami," and, *reportedly*, overnight, everyone involved was forbidden from even responding to our Retirement Village idea—ever again. *("OMG," we thought, "This [bleep-wipe] cult leader is that bad rash that won't go away.")*.

Alrighty then! We finally left for the last time! Our new mantra, "*[Bleep]* The Cult already; enough is enough!" Gotta go—***GONE***!

☞ *At the very least, we planted the seeds for the Retirement Village idea and can only hope that something positive will come out of it one day—after the pandemic.*

• • •

However, we must conclude the current chapter with this. After the passing away of Bhaktivedanta Swami, the founder and preceptor who brought the Hare Krishna Movement to the West, half of his followers begun arguing, branching off, and accusing that his own power-hungry disciples had poisoned and killed him for power.[128]

Draw your own conclusions about "sanctity" and all the *collective* takeaway from reading of the Puranas daily—morning and

---

128 *Horrific Allegations Speak of the Cult – http://bit.ly/WasTheGuruPoisoned*
Dozens search results on allegations of killing their own Guru for power and control

evening—at explicit face value. Collectively this cult seem to fail the *Puranas*!

"*[WTF!]*," I finally concluded—there was no way to help a cult trapped in their own "Motivated Reasoning."[129] All these horror stories were more than just telling! Such a preponderance of Godless mentality was evidence of the group's gross misinterpretations of the Puranas!

Do mantras work?

Personally, I finally had to leave for good after several attempts at helping and trying to find some true spirituality there. However, I needed a *sign*! I had invested a lot in bringing some clarity for myself and *service* to innocent friends and acquaintances over time. "How do I finally call it quits and fully distance from the vacuum of the 'Hare Krishna Temples?'" *I asked the sky above.* After all, I still had faith in the wisdom of the Puranas. In many forms of Hinduism, the numbers 108 and 1008[130] are significant as *"auspicious."* Rosaries or chanting beads *(mala)* typically hold 108 beads. Donations are often in some form 108 or **1008** (in Judaism, the auspicious number is 18). In any case, I elicited a meditation

---

129   *Motivated Reasoning – http://bit.ly/Motivated-Reasoning*
      Psychology Today
130   *8, 18, 108 and 1008 – http://bit.ly/108-and-1008*
      Tamil Brahmins : Hindu's Magic Numbers

approach. I sat in a *Quantum Jump Meditation*[131] and recited the Gayatri Mantra **1008** times to *Vishnu* while envisioning an astral projection in an alternate universe. ***It was just a meditation—nothing more!*** The goal was asking the Universe and all their *shaktis* for guidance if it was time to finally distance from all *The Cult's* "**H**are **K**rishna **T**emples."

That week, an envelope arrived from the New York Department of Motor Vehicles with our new registration plates. ***Confirmed!*** It was time to leave that herd behind for good and explore this *Quantum-like Science of Sambandha **on our own***.

Whether the Universe replied, or this was some angel (*shakti*) or

---

131   *Quantum Jumping – I actually recommend it! – http://bit.ly/Quantum-Jumps*
       <u>*Burt Goldman – Quantum Jumping*</u>

just dauntingly *incredible* coincidence becomes immaterial in the face of an answer to a sincere question.

But wait, many years later, during the writing of this book, after another meditation asking the Universe if I should stop, rather than continue adding to this manuscript, our Apple Books Vendor ID came in: *10081605008*! Call us "Superstitious.

Sarasvati in Park   **132**
(Illustration by the author, James Ordonez from photos on WikiMedia Commons)

- - - - - - - - - - - - - - - - -

132   *Sarasvati Statue in Park – http://bit.ly/SaraswatiParkStatue*
       WikiMediaCommons

Chapter 11

Truly Being Human
The True Magic of Yoga

"If You're Not Outraged, You're Not Paying Attention!"
~ Heather Heyer
(Hero Activist – May 29, 1985 - August 12, 2017)

W ords matter. Yes! But it is, actually, as human to fear, as it is to be angered, or to hate, as it is to love, to have sex, human contact, food, and sleep, as it is to express and communicate. Of course, it is. Humanity may have more control over our natural *creature-release needs*[133] than most animal kingdoms, by dint of our naturally advanced genetics. Yet, we are still another creature in the animal kingdom. All the above—*including hate or rage*—are still as much part of our human makeup *as it is to forgive.*

-----

133   *Creature-release needs – hunger, sleep, sex, flatulence, etc.*

By no means do we mean to suggest in this chapter that forgiveness is not paramount to everything else—of course it is. And Jesus said, *"Hate the sin and not the sinner."* Absolutely! How easy is that, really? And *Situation Ethics* gives us some basic math for the balancing act required, *"Find the most loving solution for everyone involved—always."* We can only try, and try, and continue trying.

> *One Man's Food is another Man's Poison."*
> ~*Lucretius (99 B.C.)*

We all know being perfect in a runaway world is not some auto-pilot phenomenon that should be expected from anyone. *(Unless one is a bull-headed, brainwashed cult member.)* To be alive is a complicated issue. We are blinded internally by our own human needs, our *Creature-Releases*, and the natural pains of the body, mind, and the world around us. Our inherent chemistry, emotions, and expectations tend to keep us self-absorbed—bordering on narcissism; indeed, each one of us is selfish to some degree. It's natural. This inescapable self-centeredness is another layer of *Maya's* many veils, the illusory energy of material existence.

In our takeaway, the *Puranic* suggestion is that all things in Material Existence are dialectical circumstances given substance by consciousness—*karma (action)*. Therefore, the Puranas explained *ahankara* and the cascading formation of *substances*, which subjectively encapsulate each *jiva-atman* (*living being*). Free-Will and independence are subjectivity. We, therefore, reason with our self-interest first; it's natural. "Motivated reasoning"[134] is not objective. We are motivated by that reasoning that upholds our tribal connections, be they politics, sports, religion, or some *[bleeping]* insane conspiracy theory, no matter how unfounded. Perception and belief are selective—the karmic magic behind the madness. Truth, like beauty, is *in the eye of the beholder*.

Then there are external influences from society and others. It's fair to say, from the point of view of the Puranas, another *major* blindfold that veils humanity from *mystic reality* are the myriad of values and value systems from societies and cultures that engulf us *dialectically*, in fact, all around. We sometimes hear, *"'Hate' is such a strong word, don't say you hate, say you're very upset, or disagree strongly."* Have you ever had that *[bull-bleep]* nonsense thrown at you? *"Don't use that word, don't use this word, or that."* ◌*Not allowed!*◌ We're also not allowed to be *outraged, or angry, or criticize*—in any situation; there will always be a judgment or penalty imposed by one side or another. If we show anger, outrage, or disdain, there is 'something wrong with us'—

----

134   *Motivated Reasoning – http://bit.ly/Motivated-Reasoning*
       Psychology Today

unless you don't show it, and then you can be as explosive as you [bleeping] want—in your own personal unhealthy vacuum. *"Just don't show it to anyone, and you'll be just fine."*

or...

When we see the genuinely hateful evil villain (the 'bad guy') in a movie falling to his burning death, we feel a sense of relief, justice, safe, even victory. *Actually*, we kind of really hated the *mother-[bleeper]*, as we watched! *Emotions raged.* In fact, we were almost happy they died a deservingly painful demise, *"DIE YOU [bleeping] [bleepholes]!!"* We *SUFFER*, our wrenching emotions struggling within, *"Justice and the patriotic way!!"*

Action movies—fiction or not—do move us. We laugh, we love, anger, cry, hate, and suffer until the end of the reel. Then, we tone it down just a notch—*or five*—when discussing the movie and our *"negative emotions"* with others.

Why??

In real life, however, we're not always allowed to say, *"I hate this one or that one, or Hitler, or Putin, or Stalin, or Mao,"* or the *[bleep-hole] at a distance who preys upon the innocent children.* We can think the word *"hate"* as long as we don't announce it, especially with emotion. We've tested the theory, and there is always someone in the room who believes emotions are some on-off switch. Or better yet, someone may even be thinking it's best to *bottle in all those "destructive" emotions* to keep the air clean from "negative sounds" and thoughts from others. It does not have to be Hitler or Stalin that we hate for this example, or the pedophile. Think of the last American "politician" who locked up children in cages, separated them from their mothers forever, and whose narcissist neglect statistically killed tens of thousands of pandemic patients—and all those who directly supported him in the *murderous Gross Negligence.*[135] Now imagine expressing your true feelings and emotions to several others in a *mixed* crowd. It's likely not going to go well. You will have to use filters. Scan the room. Who is going to get offended? Who is going to judge your words? Who is going to get upset, lecture you, get up and leave? The push-back typically includes those close to you, family, friends, and colleagues. Better not say anything and keep the air clean, right??

Wrong!!

Popular, sometimes unconscious-bias opinions often imply that emotions are a sign of weakness, affecting them a lack of decorum, expressing them a lack of civility, and conclusions from

---

135   *The Atlantic – Trump's Conduct Amounts to Negligent Homicide*
        <u>*Trump administration neglect kills tens of thousands:*</u>

emotional intelligence inconclusive. What matters, instead, is "strength, self-control, image, and professionalism." Never mind your own humanity and being true to yourself. "You must play" the *[bleeping]* game, or you will not get ahead, you will fail, you will be shunned from the job, cult, or circle of friends.

We judge others, and ourselves, for being human. Civility becomes the *measuring* rod; expressing outrage is the daemon, tolerance becomes the expectation, ignorance becomes bliss. The *illusion* of normality, acceptance, tolerance, even stability becomes the "*measure.*" We become compelled to fit a mold that compares to others' achievements and fitting-in. We must always be just a little more "stable" than others, in that regard, or at least a few others—being able to *measure* "stability" and "normal" is somehow urgent. All these change within each culture, religious or political cult, or tribal setting.

Let's face it, in some circles we are not allowed to admit that we experience certain emotions. Others can't handle it. Some are just inconvenienced, while others become deeply affected by another's emotional expression.

Emotions and "*Creature releases*" like hate, fear, and sexual impulses are natural yet need to be controlled — NOT suppressed. Forbidding humans to feel is both counterproductive and abusive.

Arguably, the healthiest and most holistic way to manage emotions without acting on them or suppressing them is to detach and leave situations or toxic, neglectful, or abusive individuals in the distance and move on into the infinity of the world around us. There are eight billion people on the planet.

We are taught to "face our fears and deal with them" in order to conquer the emotion. But when it comes to anger, outrage, and hate, we are told those are forbidden, *taboo*, and not human. Most psychologists, priests, and counselors refuse to deal with the issues; those human emotions are "simply not allowed." The takeaway answer from this study of the Puranas and the dialectical nature of our existence is *detachment* (*vairagya*)—creating distance and separation from toxic and non-supportive situations and *[bleepholes]*.

Sometimes a series of events eventually lead a person to the proverbial "last straw," where they realize "this is not for me." At that point, we need to detach, turn away, and end or limit relationships—even family at times. My partner and I were put to the challenge even with some family members in 2020 as we were wrapping up this book. We were smack in the middle of the Lock-Down Quarantine from the COVID-19 Global Pandemic that killed Millions. We were *"Long-Haulers"*[136] (with false-negative test results[137]) in distress because we did not get the final diagnosis until a year later, in 2021. Yet, we had all the physically painful symptoms all along. Locked-Down, literally afraid to go outside thinking we could get worse or infect someone.

The psychology was maddening with depression, anxiety, and fear of dying or infecting others, almost suicidal. The notion of going to a hospital and being intubated was the most frightening (due to the *[bleeping]* neglectful Trump administration of 2020). We were, in fact, "getting our affairs in order." Yet, some close to us did not believe we were ill, one asking for proof, as in a Doctor's Note. Others were in pandemic denial, jet-setting around on Airplanes in a complete dismissal of a *"Global Pandemic,"* despite Government health experts forbidding it. Without going into details, a "Last-Straw" came down, by "loved ones." We were being forced to violate our Quarantine against all our fears and sensibilities. The issue was over a *[bleeping]* shared parking spot in the middle of a Global Pandemic. Chronically ill, we had to figure out how to sell or get rid of our car without going outside. No one wanted to understand or believe that we were ill and

---

136   *COVID Long Haulers – https://bit.ly/Long-Haul-COVID*
        Integris Health
137   *COVID False Negatives – https://bit.ly/COVID-False-Negatives*
        PBS – 12% COVID Patients Tested Negative

fearful—throwing judgment, neglect, and hate, even expressing inconvenience over our health issues—*in the midst of a deadly genocidal global pandemic.*

That was a proverbial *"last straw."* The clear, healthy answer to any "last straws" is always to detach and distance. Distance and detachment are enunciated in Buddhism, Hinduism, and other ancient Eastern sciences. In many teachings, the Sanskrit word, *vairagya*, encourages separation, but the practice needs to be correctly understood and balanced. And indeed, the infinite vastness and dimensions of ancient and modern Quantum thought indicate entanglement through association. However, simple physical distancing is not sufficient. Forgiveness is best encouraged within the heart after disrespect, abuse, or neglect has ceased. Remaining entangled in toxic or unhealthy situations is, in simple math, not conducive to forgiveness—certainly not impossible—but by remaining we tolerate, thereby encouraging.

> That last straw dictated,
> "It was time to [bleeping] detach."
> We [bleeping] detached!

Understanding detachment is critical. We get entangled. Like the fly in the spider's web, everything and everyone we touch share resonance and dialectical intent. Cultural expectations, dogmas, and values hold us hostage—subliminally. We receive these resonances and intentions from everyone and everything around us. A *Karmic Entanglement* manifests with each interaction. The Ancient Texts allude that we indeed pick up others' *karma* by mere association. *"Don't pick up hitchhikers; you will get their karma,"* Bhaktivendata Swami would warn. We get both affected and infected by the discussed *Hegelian dialectical cascade* of ideas, ideologies, and repercussions of those around us—*"thesis, antithesis, synthesis."* Their feelings towards us, and our feelings towards them, are among the *syntheses* of any association. The effect which follows any association is a *[bleeping]* Karmic Entanglement. Likes or dislikes, love or hate, attraction or repulsion, ideologies or conclusions, intentions, and the follow-thru actions, are units of measure—*Quanta* (*Maya*).

The correlations cannot be overstated. So we again recap to dispel the illusive. The word *Quantum* means *"Measure"* (from the Latin "how much"), and *Maya* means *"Measure"* (in its foremost Sanskrit definition). Quantum Science alludes that strings are particles, and particles are strings, and both are plenums or "fields." In the Puranas, *shaktis* are energetic plenums in the foundation of Creation.

Both perspectives allude that *"Measure"* is quintessential. Inferring the material expanse is but the reductionism of immeasurable infinity into *Quantifiable* bits and pieces of the unmeasurable. And, much like the elusive electron that only reacts when being measured or the photon that only exists when observed, sound does not exist in the forest when a tree falls if there is no eardrum to measure and observe. Everything is Measure—*Quanta* (*Maya*), as are our love and hate, attraction and repulsion, inclusion and racism, relative truths and lies, conspiracy theories, ideologies, assertions, partisanships, tribal affiliations, even family—"*how much can you do for me?!*"

So we detach. It is easier said than done, but it is the *secret to forgiveness* and freedom from *Karmic Entanglement*. Like Quantum Entangled particles across great "Measurable" distances, sparks of Consciousness—the *jiva-atmans*—entangle through the association of *measure* and intention. Each association is the *thesis*, measuring into the *antithesis* of emotions, resulting in a *synthesis* of choices and actions, and so forth. The entanglement can be materially healthy or toxic;' in either case, the Science of Karma from the *Puranas* alludes to an unescapable shackle from the material expanse—not being able to return to our primordial fork in the road—beyond the limiting "*Measure*" of *Maya* or *Quanta*.

The inclusive *conclusion* of the *Puranas*—not necessarily Quantum Science—is such freedom from *Karmic Entanglement*.

The living spark (*jiva-atman*) will no longer require the encapsulation of *the 24 elemental substances (aura)* or yet another physical bodily incorporation and is free to join *Infinity*— *"Reality the Beautiful."* That, in the proverbial nutshell, is *The Magic of Karma*, from the Puranas

However, this again begins to hover on the complex topic of theosophy and religion we are trying to avoid. These are crazy times! Religions from the past may give solace, but not without learning to *detach* and stay clear of the incessant flow of karmic entanglement in today's world. Never before in the history of humanity has the situation been this dire. From overpopulation to Climate Change crises, 4,300 religions, dangerous conspiracy theories, woven political ideologies, and constant wars moving around the planet, the sense of desperation grows into an unfettered evil engulfing the planet—mimicking an "End of Days" scenario.

It's easy to be scared and want to seek shelter in a group. As the references and sciences continue to hint in previous chapters throughout this book, *detachment* from karmic entanglements appears to be necessary for individual *self-help*. Our experiences and so many others with religious and political cults are telling and guiding.

Learn to identify and stay away from cults. That, above all, is the main lesson from our experiences in this study of both ancient and modern sciences. Hypocrisy and required loyalties are the first signs to run in the opposite direction as fast as you can. Follow the wisdom of the sciences, old and new, not the sentiment of religion or ideology leaders. The confluence and similarities of old and new reasoning are telling and spookily timely as if a covenant agreement were put in place, waiting to be discovered at the right time—our takeaway for navigating the challenging future. Understand the dialectical nature of things, people, and tribal mentalities. Develop spiritual mindfulness that is neither religious nor tied to a cult. Learn from the past.

Be aware of the sciences and their development through past centuries without getting sucked into ideologies. Be progressive and involved with your voice, pen, and voting power, staying clear of old-world "*conservation*" of cultures. If we learned anything in this study and the last hundred years, the evidential preponderance is that everything and all our realities are *dialectical* transformations of *Consciousness*, ephemeral and fleeting. Getting entangled in group ideologies, churches, or group political theories is the herding instinct taking over your individuality. Stay aware. Detach from groups, people, and forces that do not support and respect your individuality and psychological makeup.

The herding instinct always tries to make us fit in and change our vibration and destinies. The future is yours to navigate and no one else's. These are new times. The past is gone. One hundred years ago, there were One Billion people on the planet; there are now Eight Billion. The earth is burning from consumption. Your future is yours to manage your destinies.

However, even without the Ancient Texts or Quantum science, there exists a Common Sense to tap into if we just look at the world and pay attention. The only prerequisite to seeing is desire. As soon as we arrive at agreeing we are not these bodies, the lens appears, and we begin to seek and also introspect. The *"Third Eye"* begins to open. Yet still, even without this new Third Eye lens, a sensical person can look at religious or political cults and identify their insidious and dangerous natures. This book was intended to help open some eyes and examine the many nuances and perspectives that point to the obvious.

Our experiences, yours and mine, already give us enough clues from our life experiences, as with recorded history. Indeed, the stories and examples of toxicity we share in this book alone speak loudly. The Twenty-first Century is a pivot, a turning point. Young people and future generations now require a new lens, perspective, and a *new Education System*. The new game is afoot. Survival requires a new navigation chart. The Old World and conservatism embracing the old have to be left behind—climate, population, and desperation alone scream this louder than ever before. Our advice is all over this book, to *detach from cults* of all types. Think for yourselves, and stop depending on leaders by accepting Gurus or blindly following politicians—they are the same *[bleep-holes]* in different dress. Vote you must! Whatever excuse one has not to vote in your country's midterms, and general elections, studying the issues is your surrender to the beast engulfing the planet. We become complicit by not being involved in that responsibility. The future is yours; you must learn to see through this new lens to navigate the obstacle course. Study these subjects from Ancient and modern sciences and take charge of the planet and your lives, so opportunistic ideology-cult leaders do not.

☞ *Getting entangled in group ideologies, churches,*
or group political theories is the herding instinct
taking over your individuality..

...

We mentioned *"celibate sex"* in earlier chapters. You guessed it; the hypocrisy is real and rampant, but not only with celibate monks and priests going after young children. Among the many scandals with religious groups and organized religions, especially

"The Cult" in question (our poster-child), are sex scandals. There are more discreet sexual relations in these celibate cults than at the local social club or corner bar. From the many Puranic allusions we can stitch together, and the Puranic explanations of the illusory nature, *Maya*, taboos, and dogmas feed the dark side within us all. Like it or not.

Secretly, insidiously, and even often subconsciously, the proverbial forbidden fruits are often-times more tantalizing, even hypnotic. We may sometimes jokingly hear, "food tastes better when you steal it" or "sex is best when dangerous, or strange and risky." Among the references in scriptural parables is the Adam and Eve story; the apple of a particular tree was "forbidden," therefore irresistible. *(And, of course, the [bleeping] patriarchs had to blame the female. Because, likely in their subconscious, she was the forbidden fruit.)* Nevertheless, our "nature," *acquired-dharma*, is stronger than all of us. Unless you're Sister Teresa or Gandhi— and who really knows what they were up to behind closed doors.

Please don't get a visual!

The story of the fox and the scorpion illustrate the point. A fox, about to swim across a river, is approached by a scorpion hitching a ride. *"No [BLEEPING] way!! You'll sting me,"* shouts the fox. The scorpion argues, *"of course I won't! If I do, we'll both drown!"*

The fox agreed, and sure enough, halfway across, the scorpion digs in. *"What the [BLEEP]? Now you're going to drown with me!"* screams the fox. *"Sorry, it's my nature,"* cried the scorpion.

The forbidden idea of celibacy and having to hide human urges makes the hormones rage. Secrecy and the danger of being found out behind it all exponentially multiply the endorphins in the air from everyone. The atmosphere becomes surcharged—an insidious incestuous undercurrent of repulsion and attraction surfaces beneath the social fabric. Every glance resembles a seduction, even when not. Urges and allegations fester, attractions turn to repulse, affairs and unlikely encounters emerge in the shadows—never to see light. The most staunch, disturbed by it all, become angry and judgmental. Every inch of skin elicits feelings—attraction or repulsion. As one dear lady friend once pointed out, *"even an elbow is a turn-on in this climate."* Swamis and monks have been caught playing with themselves, each other, their secretaries, or the altar-boys. Women cheat on their husbands much more in those climates than outside the cults. The whole *[bleeping]* thing becomes a *love-hate* smorgasbord of "celibate sex."

Angers, and tempers, and judgments, ...oh my!

And the entire social structure crumbles because of one stupid unrealistic *[bleeping]* taboo. Call us crazy—we've seen it happen to tell the story. A bunch of sexually frustrated hippies put on robes, and as if by magic, the most potent force in the universe, *procreation*, is defeated.

[Alrighty then!]

We don't need to read the Puranas to know about the "birds and the bees" and the simple "facts of life." Our genetic makeup and the modes of nature, the forces of creation, our *acquired dharma (our inherent acquired nature)* force us to attempt procreation (sex) even against our will. Those who deny this truth are clearly missing a spark-plug or two. It's simple-science and everyday observation with humankind.

> "...by what is one impelled to sinful acts (bad karma), even unwillingly, as if engaged by force?" (asked Arjuna)

> "It is lust only, Oh Arjuna, which is born of contact with the material modes of passion and later transformed into wrath, and which is the all-devouring, sinful (karmic) enemy of this world "
> ~ *Bhagavad Gita Purana 3:36-37*

Yet, "*The cult*" leaders, led by "*Mr. Swami*,", like the Catholic Church, continue in denial and refuse to change failed traditions. At the cost of many innocent children, their patriarch misinterpretations and taboos do not move them to re-examine and change course. The *"fundamentalist"* knuckleheads are stuck on a chalk line of tradition, like conservatives, with the hypnotic catatonic fear in certain animals. E.g., *the chicken goes into catatonic paralysis when confronted with a straight chalk line pointing forward.*[138] Progress is the main taboo, both with religious and political cults.

The old world cult mentality is not limited to religious and political cults. Entire societies, cultures, global educational systems, and institutions are victim to the veils of *Maya*. There exist no lack of examples. This human race is one wacky, indoctrinated herd. For instance, when we first encountered the Human Rights debacle of *Pro-Life and Pro-Choice* in our teens, back in high school, we began to ask about turning the tables with *Vasectomies*—it was like cursing out someone's mother. We are yet to hear from anyone, not one single voice our entire life; that known to be *"Effectively Reversible Vasectomies"*[139] could have contributed to solving the violent and abuse abortion debate problem for both sides many decades ago. Instead, we still get blank stares when we bring up the subject. No one will even commit to the discussion.

---

138   Catatonic paralyzed chicken – http://bit.ly/like-Hypnotizing-a-Chicken
       Catatonic paralyzed chicken fear of the chalk line
139   Vasectomy Reversal Highly Effective – http://bit.ly/VasectomyReversalHighlyEffective
       Vasectomy reversal highly effective, even after 15 years

The patriarchy runs so deep in our veins that even the proverbial "ten-foot pole' will not step forward.

The entire material paradigm is one big [bleeping] lie from one viewpoint or another. The dialectical nature of conscious-ness-based realities leaves too much wiggle room to land anything concrete on any topic, situation, or perspective.

Detach we must, or we go crazy with the rest of the asylum. Be true to yourself—be genuinely human so that you may be respon-sibly in control of the *acquired dharmas* that engulf us. Mistakes will happen; we can only grow. The Puranas teachings of *karma* and *sambandha* and the relativity behind it all—the meditation— give us the cohesion and a sense of sanity to maintain common sense on the journey.

Indeed, we don't need the Puranas to tell us how different all our *"realities"* are. Looking at the political, cultural, and religious landscapes in our various countries and across the world, we can see the variety of peculiarities and diverse realities we all perceive and believe to be real.

There is no doubt we are all crazy, deeply believing we see absur-dities that do not exist. From the QAnon cult in politics truly believing that "Jewish Space Lasers" are setting the fires in

California, or a cargo ship stuck in the Sues Canal for a week was Hillary Clinton transferring children for sex—an entire group believes the allegations.[140] Isn't that as *[bleeping]* rational as various religious cults starting wars and killing others because they believe a different absurd interpretation of the same book?

Among the examples is that an entire culture among several countries at war with each other for centuries continue keeping their women covered from head to toe with *Burqas* because the men *apparently* are afraid of sex with women.[141] Are they also then going after the altar-boys? Inquiring minds may want to know. Of course, they are![142] (*See Footnote: "The hypocrisy of child abuse in many Muslim countries" – The Guardian*)

> *If all this doesn't outrage us,*
> *then we're not [bleeping] paying attention that*
> *there's something very [bleeping] wrong!!*

By the *Yin Yang* reality of things, we are forced to filter our humanity; our feelings, language, expression, and behavior. In some moderate ways, these various *absurd value systems* are a

---

140   *Hillary Clinton accused by QAnon – http://bit.ly/QAnon-Suez-Canal-2021*
         http://bit.ly/QAnon-Suez-Canal-2021
141   *Burkas – http://bit.ly/Burkas*
         Wikipedia
142   *The Children Behind the Burka – http://bit.ly/Behind-The-Burka*
         The Guardian: The hypocrisy of child abuse in many Muslim countries

good thing, at least as some essentials to keep law and order, in certain cultures still living in the past. Yet, because these values and judgments grow organically and randomly from society to culture, to religion, even generations, unable to get regulated, some of them become dangerous *dogmas* and *taboos* into unknown futures where neither purpose nor relation applies. Inevitably, the emotional and psychological damage goes viral from the individual on through entire societies and cultures—it's a *karmic* madness from where there is no exit—so we tend to submit to the mayhem, tolerate, "go with the flow," and fit-in. Right?

Wrong again!!

The 'way of the world' is never healthy without searching deep within ourselves while reaching ever outward for an objective universal view of things. The *Puranas*, with their wisdom of the ages, allow us the comfort of balancing the dualistic reality of this *Yin Yang* existence. Not only understanding but submerging oneself in the *Puranic Science* of *sambandha* and the magic of *karma* is the essential medicine for the heart and soul of individuals and leaders. A *"mindfulness"* of our dialectical world surfaces, as with Dialectical Behavioral Therapy[143] after grasping this Quantum-like Puranic science of Karmic Entanglements. The

---

143   Dialectical Behavioral Therapy – http://bit.ly/Dialectical-Behavioral-Therapy
      DBT For Dummies In Fact, DBT For Dummies will help some wrap
      their brain around this illusive ancient Puranic Quantum Lens!

balance and perspective provided by the Puranic Science anoints a sense of universal love and belonging in an authentic relationship with all things. A journey begins—as the Puranas explain, a *"Journey Home."*

Again, in some absurd ways, these value systems do help maintain *behavioral standards* to keep order. We can search up and down the entire spectra of human *values*, *dogmas*, and *taboos*, only to find that no matter how inaccurate or outrageous, these all help maintain some order within cultures and tribes, religions, nations and laws. Yet, as essential as they may be in their subjective environment, these same values cloud over human expression and perspective and are only blinding and censoring in the end. This paradox is yet another aspect of *Maya's veil*, the illusory energy discussed in the Puranas, which keeps us from seeing reality as it is. Navigating human life blinded by avarice, judgments, and taboos imposed by others, from some distant past, is the fate of humanity, which precludes us (*you*) from being able to see *Reality The Beautiful!*[144]

According to the Puranas, what we see when we look at the world is indeed a series of veils and curtains projected from our biases and perceptions—Measures. Modern science theorizes the probability

---

144   *Swami B.R. Sridhar – http://bit.ly/RealityTheBeautiful*
        Search For Sri Krishna: Reality The Beautiful (book on Amazon)

of a *holographic universe*[145] projected from the subatomic. And the Puranic Ancient Science described a cascading transformation of *elemental substances* initiated by consciousness *(decoherence)*, which altogether project what we see and experience as physical matter into existence. From both perspectives, it's all a *[bleeping]* illusion based on outside forces and intentions from some universal life-force beyond physical matter. Illusioned by the glare, we take ourselves too seriously in these puny one hundred years, which is equal to zero in the eyes of infinity. Be real to yourself and be true to the simple math around us. Getting caught up in the madness is a *[bleeping]* disaster for oneself and our journey that awaits— whatever that may be.

Cults are everywhere, religious and political; social and cultural cults are *mindsets* sprouted from value systems and beliefs rarely based on common sense or even complete or current truths. Whether it's the QAnon cult of the United States Republican party in 2020, or *"The Cult Branch"* poster-child discussed as an example, or the college fraternity, the motivating force is always a *tangle of twisted truths frayed into knots of lies and deceit for one gain or another.* That is, these cult leaders manipulate apparent facts and insidiously weave them into posturing and presentations that are no longer factual as a whole—when put all

---

145   *Holographic Universe – http://bit.ly/Why-Universe-Hologram*
       YouTube Sabine Hossenfelder Why Universe Hologram

together. As per the Puranas and some modern Quantum perspectives, the discussed dialectical nature of reality explains and gives a sense of rational outlook to the madness of humankind and the building blocks of the world's diverse cults and indoctrinations. Humankind is a *[bleeping]* insane asylum—like it or not—from many points of view.

> *I've never had a bank account!" Whimperingly cries as if begging for compassion, one of many Millionaire Swamis (the "Renounced" order) at the start of each lecture while solely controlling hundreds of millions through his own devoted, dedicated 'disciple' accountants and attorneys.*
> *~ We'll give the reader three guesses whose clown theatrics these are.*

The QAnon QOP members are insane, cult members following cult leaders are insane, and the world's *[bleeping]* religions are all insane—keeping in mind that opposing groups like the Democrats are not off the hook. All political parties thrive on the elements that form a cult, *driven partisanship, required loyalty, and negotiated agreements* from motivated reasoning and selective perspectives and agendas. The posturing is always twisted tangles of rhetoric and interpretations, like lawyers spinning a case.

The world is a lie!

All the daily News channels consistently remind us of the

weaves, slants, tilts, and spins of truth creating realities around us. Political cults are more dangerous than religious cults. One prominent News Station lays it all out in this must-watch story[146] (see footnote).

Hiding in plain sight

The previous chapters explored the nature of *shaktis* and *[elemental]* substances alluded to in the Puranas, Vedas, and Upanishads, as potentially similar or even synonymous to particles and strings, yet *from* Consciousness *and of* Consciousness. We can make the assumption from the comparisons between ancient theories and modern science that our subatomic Quantum particles are intentions and impulses driven by the expansion of the *Ocean of Consciousness* described in these ancient texts. That is, not necessarily to assign an atom, electron, photon, or quark to a specific *[Puranic]* elemental substance, per se. The idea of infinite quantity may cloud that over.

However, the argument remains; combinations of strings and intentions may form or indicate a relationship between Quantum Particles and the twenty-four Puranic elemental substances. In essence, if truly *Consciousness* is 'the' universal driving force,

---

146   *CNN Jim Acosta – http://bit.ly/CNN-Example*
       CNN Jim Acosta – Chew on This–

the Hegelian Triad (*thesis, antithesis, synthesis*) is the perpetual momentum mechanism of intention, action, and result—*karma*. However, for whatever reasons, we tend to and choose to stay in the dark on progress and education—through *[bleeping]* conservative mentalities and approaches to life.

One hundred years ago, in 1923, it was still acceptable common knowledge and belief from centuries-old Copernicus and Euclidian theories that the span of the universe was limited to the size of our Milky Way Galaxy. There was no concept of infinity, the universe had an edge, and there was nothing else beyond it. Then in 1924, with the help of "non-Euclidian Geometry" (from Carl Friedrich Gauss and Bernhard Riemann), Einstein applied their "curved-space" findings to the space around us through his General Theory of Relativity, giving clarity to the force we call Gravity and the "Expanding Universe."

> *The equations of General Relativity… revealed that it was the presence of 'mass' that caused space to curve and distort."*
> ~ *Professore Jim AllKhalili*
> *(Everything and Nothing documentary – Amazon Prime)*

Most of us truly do not really know why it gets dark at night. Think about that; test the theory with passersby on a busy metropolitan city sidewalk. Very few will tell you that it's "the constant expansion of the universe. If the universe has been around 13.7

billion years, all starlight from that infinite number of stars should have gotten to us by now, whitewashing the night sky. However, because of the constant expansion of the universe and the stars beyond "the observable universe" constantly getting further and further away, their light has not reached us yet. Let's ponder this. Why does the general populace not know why it gets dark at night, an entire [bleeping] century after the discovery was made?

Likewise, all the other century-old Quantum discoveries and sciences are only known to avid readers, Quantum Thought enthusiasts and scholars, and the scientific community. Most folks outside universities and Physics curriculums only know bits and pieces of the magical Quantum world from fictional movies and TV shows. Just like everyone from an early age should know why it really gets dark at night, the Quantum findings of "decoherence" potentially being "*Consciousness-driven*" arguably should be a theory as familiar to young and old. And the argument continues; if eons of thinkers and scientists continue to conclude and agree there is one form of *Consciousness* or another in the microcosm and the macrocosm, why then is that not made available to general populations? The answer should be easy—World Religions, religious cults, and political cults—conservative greed, fear, and avarice for power.

As a society and civilization, it becomes urgent to understand and educate all emerging scientific discoveries, which holistically change our view of the world for our entire lifetimes across generations. Waiting for several generations, to tell the truth in schools is dangerous procrastination, holding back both education and social progress. When I was in third grade, we were taught Square Roots and Basic Algebra; not because it was a "special" or "gifted" class. That was just the standard curriculum in Colombia, S.A., for that grade. I arrived in the United States to a fourth-grade class stuck in Addition through Multiplication.

The dumbing-down of the masses by inaction (*akarma*) might be unintentional from the elite in power. Still, if we follow "Murphy's Law" or the "Infinite Monkey Theorem,[147]" if it can happen, it will happen.

Euclidian Geometry and classical physics before Einstein, which are still taught in schools inadvertently as some form of absolute end, still imply that gravity is a force that brings objects together. *Raise your hand if that is your understanding.* However, the hundred-year-old scientific facts left out of many school systems is that objects falling are "just following the simplest path through bent space."

---

147   Infinite Monkey Theorem – https://bit.ly/Infinite-Monkey-Theorem

> *Gravity is nothing more than the curvature of space itself; where [when] an object falls, it's not being pulled by gravity at all; it's just following the simplest path through bent space. The reason we have gravity on Earth is because the Earth is actually bending the space around it."*
> ~ *Professore Jim AllKhalili*
> *(Everything and Nothing documentary – Amazon Prime)*

The reason for the hold-up is likely that stubborn old-world conservatism based on old-world religions and fear-of-change are the stumbling blocks of progress and safety for humanity—and sadly, the safety of the planet.

Imagine if we were to upgrade education systems to allow young people to ponder the conclusions of the ancient and modern Quantum sciences, *without* religion, yet still accepting *"Consciousness"* as a scientific driving force. Could that shape future generations into better human beings, understanding and embracing basic altruistic principles without the need for religious racism, tribal cults, and separatism?

Would the scientific idea and mass cognitive awareness of a *"Consciousness"* lurking in everything, without amplifying churches' power, fear, and control, perhaps be a new start for civilization? In the Ancient Quantum like Texts, the subatomic elemental plenums and substances arguably mirror today's obser-vations of wavelengths and fields as described in modern science.

Why not place the comparison before all educational audiences to ponder, revel, and experiment? In chapter five, back at our *"Primordial Fork,"* we followed the puranic allusions of the *Jiva-Atman*, a spark of *consciousness* awakening to a decision, and a sojourn, eliciting materials—substances—to build and accommodate its trek through the material expanse. As symbolic as that may be, a "science" is described even beyond the written allegorical vernacular. The essence is a Particle Science of energetic plenums and fields, albeit dialectically, building and sculpting living bodies, habitats, and civilizations—a scientific Creationism.

As we continue bringing it all together, not only with both sciences (*Ancient and Quantum*) but also our new common sense when we use this mindful vernacular (*sambandha*) to witness the bird's eye view, a whole new reality emerges. The world is a lie, period! We need to be careful not to apply this "lie" to *only* ideologies and beliefs. The cascade of dialectical realities does not stop there—this requires some discussion:

A bridge is built, for instance, for a need to cross, which came from an idea to facilitate a previous dialectical need or desire—

mobility. The bridge is 'real,' to us, in its tactile brick, mortar, and steel construction, but much more fundamental, therefore more substantial, is the necessity for that bridge, without which there would be no bridge.

This conundrum is the colossal snare from the Karmic Illusion (*Maya*), whose extreme simplicity hides truth and reality right under our noses. The bridge, in this case, is a symptom of a higher reality—necessity, desire, thought, planning, gathering, and constructing. And we all know this simple process, yet we all completely dismiss the true *substance* behind the bridge.

Our material senses will tell us, "If you can't touch it, see it, taste it or smell it, then it is not real." Yet, as we read from Ancient Science and Quantum Thought, our material senses are indeed also mere constructs from the cascade of dialectical intentions, desires, and needs. Everything is *consciousness* and is driven by *consciousness*. Likewise, building the bridge, its materials, planning, and resources all are a series of karmic cascades (*action*), purely driven by Hegel's dialectical triad—*thesis, antithesis, synthesis*—karma *(simple 'action')*. Without human thoughts, needs, and intentions, we would have no cities, infra-structures, or societies. The Earth would only be the wilderness dialectically born for the needs of all *other* species utilizing the Earth's ecology.

The feeding and survival instinctive needs of all creatures, build river dams and bridges, as with beavers, or the many different types of colonies by bees, ants, spiders, or prairie dogs, for their *dialectical instinctive* predispositions. Some spiders will curl leaves into shelters inside their webs. Termites build colossal air-conditioned mounds. Some bird's nests, like the Rufous Hornero, build unique constructs. This Rufous Hornero, a South American bird, strategically builds earthen nests in trees, from mud and dung into coned-shaped bowls high on tree branches. The sun then hardens the nests for shelter, which face away from the prevailing winds, for refuge from the weather. Prairie dogs have been known to construct underground "towns" up to 25,000 square miles, with dedicated chambers for nurseries, food storage, and shelters from weather conditions. Trees *talk* and *share* with each other for health and growth.[148] Needs, desires, survival are all dialectical intentions that drive the creations of many types of ecosystems and infrastructures—among all species of life.

The creation of societies and cities is not limited to humankind, as the arrogant tend to dismiss. *The Magic of Karma* is just

---

148   *NPR - Trees Talk To Each Other* – https://bit.ly/TreesCommunicating

that—the magic of intentions, thoughts, needs, and desires that propose and set into motion everything in the universe. Because consciousness drives everything, the universe is alive—a living, breathing organism. This most evident of all truths—most importantly—is *"The Magic of Karma!"*

The insanity we see on the planet today with conspiracy theories, cults in politics and religion, causing growing hatred and racism, is simply desperation born of ignorance (*tamas*). The Ancient Texts (the Puranas) we discussed throughout this book speak of the "Three Modes of Material Nature—*Goodness, Passion, and Ignorance.*" These are states of consciousness wrestling with each other to facilitate action (*karma*). As our global societies dive deeper into the perils of the Nuclear Age, overpopulation, Climate Change, mass migrations, fear, misinformation, racism, and hatred, the state of *Ignorance* (*tamas*) grows like a virus, upstaging Goodness (*sattva*) and even Passion (*rajas*). The recent growing misinformation paradigm did not begin with the clown Trump or the QOP *(QAnon Republican Party)*; those are merely symptoms.

The illusion which breeds misinformation, hatred, and racism is the discussed temporary acquired *dharma* nature of eternal consciousnesses (*plural*), living beings identifying with temporary measurable existence (*Maya*).

There lies the rub!

Comically, the growing misinformation paradigm, by political *and* religious cult leaders, is the *[bleeping]* Zombie Apocalypse feared and written about since ancient times in scriptures and by prophets, portraying what a mayhem future for humankind could likely become once it got out of control. Climate Change, overpopulation, and the avarice of runaway Capitalism altogether burn an uncontainable wildfire. Like the bridge discussed, the reality and foundation of the climate catastrophe is dialectical cause-and-effect from human greed, misplaced desires and intentions, cheating, and self-aggrandizement. There is no fix because the reality is not the "Climate Reality," but the Human Reality.

It's officially out of control,
beyond the point of no return!
Experts say. [149]

---

149   *The Independent* – https://bit.ly/BeyondReturn)

The controversial idea of a new Re-Education System paradigm appears to be in order, not about Critical Race Theories, Wokeness, conformity, politics, or religion. Instead, the realities from Quantum Thought, *Karma,* and Situation Ethics—Understanding the substance of action—*"The Magic of Karma."* For all of Earth's peoples, the idea of escape from inevitable dangers ahead seems mathematically improbable in light of the preponderance of mayhem already natural to humankind. Additionally, the science of karma from the ancient Puranas like the Bhagavad Gita and Upanishads alludes to the nature of predominating *shaktis* like *Maya* (the illusory energy) prohibiting any escape from the combined forces that govern the universe—mainly due to misplaced and misapplied consciousness. Unlike animals, who don't have a developed sense of importance or narcissistic rationale, humankind creates its own demises. Therefore, *Self-Help* and Re-Education remain the only solace and shelter for any future of hope. Help is, and will only be, available to the open-minded, progressive individuals and communities willing to bend with new tomorrows.

Since unlikely the entire planet—individuals, families, and perhaps smaller organizations, schools, communities, even local governments may consider leading this proposed *Re-Education Paradigm Shift*. An upgrade is needed from superseded Classical

Physics toward Quantum Thought. And especially, education needs to move away from *Religious-Racism* taught by religious schools and religious families to preschoolers and children. That alone, by far, is the foremost evil cause of racism and hatred in today's societies; it's a *"grooming"*—like it or not—do the [*bleeping*] math! The ungodly notion unconsciously, subtly, yet insidiously fed to children that others who differ in belief, caste, culture, or dress are lesser in God's eyes is the ugly proverbial elephant in the room.

We are all guilty! God is alive everywhere in mass Consciousnesses, but Religion is always the Devil, everywhere. Above all, their widespread educational foundations underline the mass stupidity and aboriginal neanderthal nature of today's "Conservative" humanity. Education systems, even governments, are still commanded by misplaced, misinterpreted, contradicting cult and tribal religious ideologies from the past—and that is among the fundamental messages parents, and religious schools push unto our children. And there are and will be, Karmic consequences for all those responsible, including parents, teachers, and clerics—like it or not! Karmic reactions, unfortunately, by design, do not turn a blind eye to ignorance, as in any Court of Law. Imagine telling a judge, "Your Honor, I did not know this crime was illegal!"

Religions need to get in their *[bleeping]* lane and stay in their *[bleeping]* lane. They are complicit in destroying the planet! Freedom of Worship is not synonymous with Freedom of Education. Teaching children eliteness or superiority feeds the roots of racism—we do so by placing our God above others' concept of God—at an early age. No one needs a Ph.D. to do that simple math. We're confident most atheists (except, of course, atheist Trumpers and atheist members of the *[bleeping]* QAnon's QOP Republican/Communist/Fascist Party) would agree that religion has no place in Government. Governmental decisions that affect women's reproductive choices and rights, or dictate which people have more liberties to democratic voting rights, or have more freedoms than others, remain in the *religious dialectics* despite our Constitutional Law of "Separation of Church and State."[150] [151] [152] [153] ***Why is that?*** You guessed it; simple dialectical manipulation of the masses through religion for power, money, and avarice of the few—by "the 1%!" History has shown us that it is extremely easy and common practice to weaponize religion.

That insanity is true globally to one extent or another in all societies

---

150   *Freedom Forum: Church and State* – https://bit.ly/Church-n-State
151   *JSTOR* – https://bit.ly/ReligiousElectionFactor
152   *American Historians Organization* – https://bit.ly/EvangelicalPolitics
153   *Oxford Academic: Christian Nationalism* – https://bit.ly/Christian-Nationalism

due to the dialectical conservative ideology—conserving failed pasts, dismissing progress and new tomorrows.

> Is religion "the opiate of the people?"
> It appears inarguably so.[154]

Arguably, Conservative and Republican Parties of various nations have become the very Devil they "crusade" against, ironically donning the color red. As Aldous Huxley was quoted in chapter one, "Every crusader is apt to go mad. He is haunted by the wickedness which he attributes to his enemies; it becomes in some sort a part of him." The obsession for absolute control by *their* partisan ideology is akin to the Totalitarianism result of both Communism and Fascism of the 20th Century.

The obsessive dialectical crusade of lies, propaganda, disinformation, and conspiracy theories to discredit any other side, at even the cost of life-and-death at the hands of a Global Pandemic, to gain total control and hold on to failed pasts yielded the same result—*genocides*. They are as guilty as dictators, not just the politicians but also the partisan sheep who continued to fight against vaccines and *"medical-masking"* during a *proven Global Pandemic*. Karma is a *[bleeping]* bitch, and does not excuse ignorance (*tamas*).

---

154   *The Hedgehog Review* – https://bit.ly/ReligionTheProblem)

The Daily Beast article by J.P. O'Malley, "Communism and Fascism: The Reason They Are So Similar,"[155] a quote from political Scientist Robert Paxton begins to show how the tactics of the Republican Party indeed truly mirror Fascism and Fascists with awkward yet productive collaboration by massed nationalist militants, and the elite 1% class, committed to destroying democratic liberties, unethically pursuing "internal cleansing," and "external expansion." Has the GOP, *now the QOP*, become a Communist/Fascist Party? It certainly looks that way, no matter which angle we try to position it. They became what they beheld— their crusade against the Devil turned them into the same Devil. That is another example of the dialectical black, dark side of *the Magic of Karma* behind the curtain.

Still, these dangerous cult ideologies of insidious hate and separatism are not limited to partisan political cults. Our *"Poster-Child"* religious cult exemplifies such precise cult nature in organized religions.

For instance, the poster-child cult leader *RatNut Swami (Mr. Swami)* can be seen in his dealings, finances, and criminal history to mirror Trump and the Trump legacy—cent percent. He is a charismatic liar and criminal hiding behind dialectical positions

---

155   *Communism & Fascism: The Reason They Are So Similar – https://bit.ly/FascistIdeologies*
      The Daily Beast article by J.P. O'Malley

of falsehoods while convincing his cult followers that he is some *messiah*. And that is the same old story with the Taliban, Isis, Scientology, or the Catholic Church. It is always the same dialectical template; only the players and slanted or twisted ideologies change. Combining the ancient science from the Vedic Puranas with modern Quantum Thought gives us the vernacular to see *transparently* through this chaos.

☞ *Definition:* ***in·ef·fa·ble*** */inˈefəb(ə)l/ adjective*
"Too great or extreme to be expressed or described in words."

The "ineffable" magic
behind the curtain of creation

As we have done throughout the book, we use Earthly examples to illustrate the dynamic dialectical nature of karma (*simple action and its owners*). An example of the insidious dialectics from karma creating and maintaining mass political control is, for instance, the *"Dark Money" IRS 501(c)4 debacle*. **This is juicy**. Over time, politicians in the United States have stitched together a dialectical formula to run an oligarchic-fascist-totalitarian foundation, well hidden behind all the rhetoric and back-and-forth. This example is just one more lens to add clarity on how the illusions of Maya are manipulated dialectically by cults and the forces of avarice—the black magic of karma. This formula, which appears to be strategically planned and set into motion, is the aggregate locking mechanism of combined "Corporate Personhood," "Citizens United," and this IRS 501(c)4 IRS *"undisclosed"* Tax Exemption, which keeps the donors (*bribes*) absolutely secret and sealed.[156] The reader may look these up individually to understand each nuance; yet, altogether, the combination in the equation is the legalization of bribery by corporations and "the 1%" of any-and-all politicians for corporate political control, e.g., The NRA (*mass shootings*), Large Tech (*monopolies*), etc. The formula is dialectical because it is built on arguments presupposing other freedoms and considerations individually. Altogether, however, they become a fascist dictatorial oligarchy enslaving the 99% by "The 1%." And, no one is able to see this *IRS 501(c)4 legal bribery*, and no one is writing about it, and no one is campaigning against it; when brought up, it just gets as insidiously dismissed by all sides in politics just as it insidiously emerged from the shadows into a reality that governs all our lives.

Why?
Because ALL sides, ALL Parties,
ALL News Media Channels,
ALL Politicians, ALL corporations,
ALL the "1%,"
are ALL basking in the benefits
of this dialectical "Dark Money."

---

156   *N.Y.U. Journal of Legislation & Public Policy* – PDF: https://bit.ly/501c4_DarkMoney

To begin with, for example, the illusion of currency is a *dialectical* farce—a dialectical *"middleman"* in bartering goods.[157] Until 1933, the U.S. Dollar was backed up by real gold. Still, the real gold was also a middleman in ideological value because of the scarcity and demand of a pretty substance throughout history. Remember, *"Maya"* means "Measure;" for our purposes, "Measuring the unmeasurable"—a big *[bleeping]* lie, which became a reality and the foundation of an entire civilization. That is the dialectical reality of value, labor, goods, and freedom—and money buys freedom. *Are you dizzy yet?*

Understanding and being able to see the *dialectics* in all action (*karma*), with Quantum Science and ancient *Sambandha* vernaculars, through this lens, in all situations, with detachment and distancing, is by far the best *Self-Help* for the individual and the Planet.

With all the wonders being discovered, almost daily, in the universe around us, the millions of life-forms known to us on this planet alone, and the myriad of possibilities clearly still to be discovered, plus our new vernacular, it becomes exceedingly clear that *Consciousness* is everywhere—strewn across the universe and the very foundation of all universes. This math indeed points to a

---

157   *Iowa State University* - https://bit.ly/Illusion-of-Money
       Why Can't We Just Print More Money? Professor Joydeep Bhattacharya

high probability of many other types of life-forms in the universe, unlike any here on Earth, and driving dialectical karmic environments, and not necessarily carbon-based life. Science postulates to possibilities of Silicon-based life-forms in other parts of the universe,[158] and the Ancient Vedas and Puranas allude to *"ethereal,"* non-solid, subtle bodies as higher forms of intelligent life—sometimes understood as separate dimensions. Nevertheless, grasping this simplicity of a *dialectical* karmic existence (*sambandha*) is paramount to understanding the myriad possibilities and the nature of illusions and falsehoods driving societies, cultures, politics, and religious cults.

This book is the first of a series, as a foundation, towards a *Re-Education* and a support platform for discussion and collaboration with other authors and speakers to help heal our planet from the abuses of Climate Change and political control by the evil forces of avarice and greed. To avoid toxicity, racism, hatred, and even complicity, one must learn to detach with this Quantum and Ancient *Mindfulness-Vernacular* (*Sambandha*). Detachment (*vairagya* / *Kayvalia*) is the tool to maintain a healthy distance and stay clear of *Karmic Entanglements* from individuals and

---

158   NBC News – https://bit.ly/SiliconLife
       Silicon Based Life

forces caught up in false religious and political ideologies and manufactured conspiracy theories.

behind all the confusion,

Yet, God does exist and persists behind the curtain of Creation. *They* humorously wait for you to detach from *ALL* the tribes, which we all managed to knit into this tangle of reality. Detaching does not mean, as our poster-child *"Cult"* will tell you, "to surrender unto *them*." While always misinterpreted by them, the *Bhagavad Gita's conclusion* clearly states this aphorism ever so clearly. In fact, the very verse gives light to the insanity of *"The Cult"* members who believe that 'their particular interpretation' is the unmentioned exception in the same exact teaching—truly believing something is there when it's not—like a cult. The Bhagavad Gita verses clearly state:

"I am seated in everyone's heart, and from Me come remembrance, knowledge and forgetfulness... Abandon ALL varieties of religion and just surrender unto Me. I shall deliver you from all sinful reaction (bad karma). Do not fear."
~*Bhagavad Gita Purana*

The teaching there is to "Surrender to Me *(Vishnu in the heart)*," NOT the *[bleeping]* organization or leaders of the cult, whatever crimes and mischief they are allowed to commit.

Therefore, just [bleeping] Detach!!

The Puranas, among their many conclusions, implore humanity to detach. Detach from all that is toxic, situations, groups, cults or tribes, even friends and family, when respect, support, kindness, or empathy are not reciprocal or just nonexistent. Even if they just don't understand you, this life is too short and challenging to begin with, to stay connected with forces, systems, mindsets, or people whose *acquired dharma* does not support your journey.

[BLEEP] them! Detach!

At a bare minimum, a familiarity of the Puranic science of *sambandha* and *karma*, preferably total indulgence, begins to shed light and a sense of sanity to a crazy world—without surrendering to it. Oftentimes, out of pure ignorance *(tamas)*, the Puranas explain, people and the world can be cruel and unjust. Some without even knowing empirically, covered by their own unconscious bias, avarice, and narcissism—believing they are the bodies they inhabit—distinct and intimidated by the various hues of their skin, language, or places of birth—*racism*. The insanity is accidental, yet it is all around us. The Puranas explain this *Maya* as a threefold *modus operandi* of the material nature that engulfs us. These are cosmic forces from consciousness known as *Goodness (sattva), Passion (rajas), and Ignorance (tamas)*. They are not going away! These are a sort of programming based on *karma*, which we will go into in another edition. We cannot fix the world around us or others. We can only fix ourselves and lead by example—*with detachment.*

Understanding this inescapable reality becomes exceedingly crucial for leaders and do-gooders to understand the limitations and aim for realistic middle-way goals like Democratic-Socialism. Both Communism and Capitalism have consistently failed the working-class creating inequality, poverty, and segregation. Getting a sober, realistic handle on the observable fact that everything in creation is *dialectical* and ultimately based on *consciousness* may begin a healing therapy for leaders and the masses without getting all *[bleeping]* religious, sectarian, and segregationally racist.

☞ *Current political goals will never work without getting this dialectical-decoherence truth from both modern Quantum Science and the Ancient Texts! If we do not listen to the science, we will remain in the failed past.*

Conclusions?

For leaders, teachers, and influencers, the new, 100-year-late Education System paradigm is not only overdue but now highly damaging. We must cease teaching K through 12+ *only* Classical Physics; without new Quantum Thought, principles, and possibilities—side by side. Comparatively, that might be the easy part—still, good luck with that. The ugly *[bleeping]* elephant in this room is "Religious Education!" We teach our children from pre-school that God will punish others, in one manner or another, for not being "like us." And why is that still *[bleeping]* acceptable—seriously?? *There* lies the dialectical *[bleeping]* idiocy of humankind. That "freedom" is why we are not yet "Civilized." We are free to indoctrinate children into religious racism, separatism, and supremacy—yet no one dares come forward, not even the *proverbial ten-foot-pole*. While we may all understand the sensitivity and sensibility of certain "freedoms," we all agree that it is illegal, immoral, and just wrong to yell "Fire!" in a crowded theater. Why is teaching innocent children subliminal hatred acceptable by any norms, especially for priests, swamis, rabbis, and clerics? Like Freedom of Speech, all "freedoms" need to be regulated, especially "Religious Freedom." Let's grow some spines and *[cojones]* and challenge conservative traditions from a clearly failed past.

For individual *Self-Help*, stay away from all *[bleep-holes]* and cults. *Detach, detach, detach!* Detach not only from toxic people but anyone who does not support your journey or makes no concerted effort to understand you. They do not deserve the beauty of your presence in their lives. Just try to respect them from afar as *simultaneously one and different infinitesimal* sparks of consciousness from an infinite *Ocean of Consciousness*. Only then, from a distance, are many of us truly able to forgive!

Have a good *[bleeping]* life—it's a *short* chapter—prepare for the next—and don't forget, stay *[bleeping]* *detached from any toxic Karmic Entanglement from those that don't support and respect you*. It'll be the best thing you can do for the planet!

> "As grains of sand come together and are separated again by the force of ocean waves, living entities in material bodies come together and are separated again by the Force of Time."
> ~*Bhagavatam Purana 2:15:3*
>
> ...
>
> We wrote this to the sky above

		GROSS substance	TRAIT objects of the senses	ABILITY receiving sense organs	ACTIVE working senses
Gross	intelligence	earth	odor	nose	
		water	taste	tongue	
	mind	fire	form	eyes	
		air	feel	skin	
	ahankara [ego]	ether	sound	ears	
Subtle	jiva atma				

S U B S T A N C E S

©2021 James T. Ordoñez The Magic of Karma

Sambandha – The Dialectical Transformations of Substances (The Magic of Karma)

All these **substances** are strewn everywhere unmanifest, (pradhana)
until touched by Time, due to the desire of the jiva-atman.
(Illustration by the author, James Ordonez)

. . .

in infinity our sparks may meet

The End

...

...is closer than you think

...Oh My!
Did you, and us, just "stitch together"
another [bleeping] conspiracy
theory or cult??

www.ingramcontent.com/pod-product-compliance
Lightning Source LLC
Chambersburg PA
CBHW051820040426
42447CB00006B/295